FM 3-0

OPERATIONS

FEBRUARY 2008

HEADQUARTERS
DEPARTMENT OF THE ARMY

Published by Books Express Publishing
Copyright © Books Express, 2010
ISBN 978-1-907521-35-5
To purchase copies at discounted prices please contact
info@books-express.com

Foreword

America is at war, and we live in a world where global terrorism and extremist ideologies are realities. The Army has analytically looked at the future, and we believe our Nation will continue to be engaged in an era of **"persistent conflict"—a period of protracted confrontation among states, nonstate, and individual actors increasingly willing to use violence to achieve their political and ideological ends.**

The operational environment in which this persistent conflict will be waged will be complex, multidimensional, and increasingly fought "among the people." Previously, we sought to separate people from the battlefield so that we could engage and destroy enemies and seize terrain. While we recognize our enduring requirement to fight and win, we also recognize that people are frequently part of the terrain and their support is a principal determinant of success in future conflicts.

This edition of FM 3-0, the first update since September 11, 2001, is a revolutionary departure from past doctrine. It describes an operational concept where **commanders employ offensive, defensive, and stability or civil support operations simultaneously as part of an interdependent joint force to seize, retain, and exploit the initiative, accepting prudent risk to create opportunities to achieve decisive results**. Just as the 1976 edition of FM 100-5 began to take the Army from the rice paddies of Vietnam to the battlefield of Western Europe, this edition will take us into the 21st century urban battlefields among the people without losing our capabilities to dominate the higher conventional end of the spectrum of conflict.

Ours is a doctrinally-based Army. FM 3-0 provides the intellectual underpinnings that lie at the core of how our Army will organize, train, equip, and conduct operations in this new environment. It recognizes that **we will achieve victory in this changed environment of persistent conflict only by conducting military operations in concert with diplomatic, informational, and economic efforts.** Battlefield success is no longer enough; final victory requires concurrent stability operations to lay the foundation for lasting peace.

Although the strategic environment and operational concepts have changed, **Soldiers remain the centerpiece and foundation of the Army—as they have been since 1775.** These Soldiers are led by leaders proficient in their core competencies, sufficiently broad to adapt to conditions across the spectrum of conflict, and courageous enough to see enemy vulnerabilities and exploit opportunities in the challenges and complexities of our operating environments. As leaders, it is our obligation to understand and be proficient at employing Soldiers in full spectrum operations. We must read, study, understand, and implement the doctrine in FM 3-0.

WILLIAM S. WALLACE
General, U.S. Army
Commander
U.S. Army Training and Doctrine Command

Field Manual
No. 3-0

Headquarters
Department of the Army
Washington, DC, 27 February 2008

OPERATIONS

Contents

Distribution Restriction: Approved for public release; distribution is unlimited.

*This publication supersedes FM 3-0, 14 June 2001.

Figures

Tables

Preface

FM 3-0 is one of the Army's two capstone doctrinal publications; the other is FM 1, *The Army*. FM 3-0 presents overarching doctrinal guidance and direction for conducting operations. The eight chapters that make up this edition of *Operations* constitute the Army's view of how it conducts prompt and sustained operations on land and sets the foundation for developing the other fundamentals and tactics, techniques, and procedures detailed in subordinate field manuals. FM 3-0 also provides operational guidance for commanders and trainers at all echelons and forms the foundation for Army Education System curricula:

- Chapter 1 establishes the context of land operations in terms of a global environment of persistent conflict, the operational environment, and unified action. It discusses the Army's expeditionary and campaign capabilities while emphasizing that it is Soldiers and leaders who remain the Army's most important advantage.

- Chapter 2 describes a spectrum of conflict extending from stable peace to general war. From that spectrum, it establishes five operational themes into which various joint operations fit. This chapter helps Army leaders to understand and differentiate between the requirements of diverse joint operations such as peacekeeping and counterinsurgency. It shapes supporting doctrine for each operational theme.

- Chapter 3 is the most important chapter in the book; it describes the Army's operational concept—full spectrum operations. Full spectrum operations seize, retain, and exploit the initiative and achieve decisive results through combinations of four elements: offense, defense, and stability or civil support operations. It establishes mission command as the preferred method of exercising battle command.

- Chapter 4 addresses combat power, the means by which Army forces conduct full spectrum operations. It replaces the older battlefield operating systems ("BOS") with six warfighting functions, bound by leadership and employing information as the elements of combat power. Combined arms and mutual support are the payoff.

- Chapter 5 reviews the principles of command and control and how they affect the operations process—plan, prepare, execute, and assess. The emphasis is on commanders and the central role that they have in battle command. Commanders understand, visualize, describe, direct, lead, and continually assess.

- Chapter 6 discusses operational art, offering Army commanders a bridge between military theory and practice.

- Chapter 7 is about information superiority, particularly the five Army information tasks, purpose, and staff responsibility.

- Chapter 8 discusses the requirement for Army forces in joint campaigns conducted across intercontinental distances. It frames the challenges created by the requirement for Army forces in terms of strategic and operational reach.

Four appendixes complement the body of the manual. The principles of war and operations are in appendix A. Command and support relationships are in appendix B. A brief description of modular force is in appendix C. And a discussion of the purpose of doctrine in the Army is in appendix D. This appendix includes a chapter-by-chapter summary of the important changes made in this edition of FM 3-0. It also includes tables listing new, modified, and rescinded terms for which this manual is the proponent.

Army doctrine is consistent and compatible with joint doctrine. FM 3-0 links landpower doctrine to joint operations doctrine as expressed in joint doctrinal publications, specifically, JP 3-0, *Doctrine for Joint Operations*. FM 3-0 also uses text and concepts developed with North Atlantic Treaty Organization partners. When published, Allied Joint Publication 3.2, *Doctrine for Land Operations*, will contain this material.

The principal audience for FM 3-0 is the middle and senior leadership of the Army, officers in the rank of major and above who command Army forces in major operations and campaigns or serve on the staffs that support those commanders. It is also applicable to the civilian leadership of the Army.

FM 3-0 uses joint terms where applicable. Most terms with joint or Army definitions are in both the glossary and the text. *Glossary references*: Terms for which FM 3-0 is the proponent publication (the authority) have an asterisk in the glossary. *Text references*: Definitions for which FM 3-0 is the proponent publication are in boldfaced text. These terms and their definitions will be in the next revision of FM 1-02. For other definitions in the text, the term is italicized and the number of the proponent publication follows the definition.

"Adversaries" refers to both enemies and adversaries when used in joint definitions.

"Opponents" refers to enemies and adversaries.

FM 3-0 applies to the Active Army, Army National Guard/Army National Guard of the United States, and U.S. Army Reserve unless otherwise stated.

This manual contains copyrighted material.

Headquarters, U.S. Army Training and Doctrine Command, is the proponent for this publication. The preparing agency is the Combined Arms Doctrine Directorate, U.S. Army Combined Arms Center. Send written comments and recommendations on a DA Form 2028 (Recommended Changes to Publications and Blank Forms) to Commander, U.S. Army Combined Arms Center and Fort Leavenworth, ATTN: ATZL-CD (FM 3-0), 201 Reynolds Avenue, Fort Leavenworth, KS 66027-2337; by e-mail to leav-cadd-web-cadd@conus.army.mil; or submit an electronic DA Form 2028.

Acknowledgments

The copyright owners listed here have granted permission to reproduce material from their works. Other sources of quotations are listed in the source notes.

On War, by Carl von Clausewitz, edited and translated by Michael Howard and Peter Paret. Reproduced with permission of Princeton University Press. Copyright © 1984.

Introduction

This is the fifteenth edition of the Army's capstone operations manual. Its lineage goes back to the first doctrine written for the new American Army, Baron von Steuben's 1779 *Regulations for the Order and Discipline of the Troops of the United States*. Today, as with each previous version of *Operations*, FM 3-0 shapes all of Army doctrine, while influencing the Army's organization, training, materiel, leadership and education, and Soldier concerns. But its contents are not truly capstone doctrine until Army forces internalize it. This requires education and individual study by all Army leaders. And it requires more: Army leaders must examine and debate the doctrine, measuring it against their experience and strategic, operational, and tactical realities. They must also recognize that while FM 3-0 can inform them of how to think about operations, it cannot provide a recipe for what to do on the battlefield.

Always dynamic, Army doctrine balances between the Army's current capabilities and situation with its projected requirements for future operations. At the same time, Army doctrine forecasts the immediate future in terms of organizational, intellectual, and technological developments. This requirement is particularly challenging for this edition of FM 3-0. The Army is heavily committed in conflicts in Afghanistan and Iraq and to countering terrorism worldwide. How long this will remain the case remains unknown. Therefore, this edition promulgates doctrine for Army operations in those conflicts. However, America's strategic requirements remain global. FM 3-0 does not focus exclusively on current operations, regardless of how pressing their requirements. The Army's experience makes it clear that no one can accurately predict the nature, location, or duration of the next conflict. So this doctrine also addresses the needs of an Army responsible for deploying forces promptly at any time, in any environment, against any adversary. This is its expeditionary capability. Once deployed, the Army operates for extended periods across the spectrum of conflict, from stable peace through general war. This is its campaign capability.

This edition of FM 3-0 reflects Army thinking in a complex period of prolonged conflicts and opportunities. The doctrine recognizes that current conflicts defy solution by military means alone and that landpower, while critical, is only part of each campaign. Success in future conflicts will require the protracted application of all the instruments of national power—diplomatic, informational, military, and economic. Because of this, Army doctrine now equally weights tasks dealing with the population—stability or civil support—with those related to offensive and defensive operations. This parity is critical; it recognizes that 21st century conflict involves more than combat between armed opponents. While defeating the enemy with offensive and defensive operations, Army forces simultaneously shape the broader situation through nonlethal actions to restore security and normalcy to the local populace.

Soldiers operate among populations, not adjacent to them or above them. They often face the enemy among noncombatants, with little to distinguish one from the other until combat erupts. Killing or capturing the enemy in proximity to noncombatants complicates land operations exponentially. Winning battles and engagements is important but alone is not sufficient. Shaping the civil situation is just as important to success. Informing the public and influencing specific audiences is central to mission accomplishment. Within the context of current operations worldwide, stability operations are often as important as—or more important than—offensive and defensive operations. Department of Defense policy states:

> *Stability operations are a core U.S. military mission that the Department of Defense shall be prepared to conduct and support. They shall be given priority comparable to combat operations and be explicitly addressed and integrated across all DOD activities including doctrine, organizations, training, education, exercises, materiel, leadership, personnel, facilities, and planning.*
>
> DODD 3000.05

Because of this, full spectrum operations—simultaneous offensive, defensive, and stability or civil support operations—is the primary theme of this manual. This continues a major shift in Army doctrine that began with FM 3-0 (2001) and now is embedded in joint doctrine as well. Stability and civil support operations cannot be something that the Army conducts in "other than war" operations. Army forces must address the civil situation directly and continuously, combining tactical tasks directed at noncombatants with tactical tasks directed against the enemy. These tasks have evolved from specialized ancillary activities—civil-military operations—into a central element of operations equal in importance to the offense and defense—stability and civil support. The nature of the mission determines the appropriate weighting and combination of tasks.

The emergence of full spectrum operations drives key changes in capstone doctrine. The Army established full spectrum operations in FM 3-0 (2001), shifting sharply from an "either-or" view of combat and other operations to an inclusive doctrine that emphasized the essentiality of nonlethal actions with combat actions. This edition of FM 3-0 continues that development. In FM 3-0 (2001), stability operations were "other" joint missions stated in an Army context. The current edition describes stability operations as tactical tasks applicable at all echelons of Army forces deployed outside the United States. In addition, civil support operations are also defined as tactical-level tasks, similar to stability tasks but conducted in the very different operational environment of the United States and its territories.

The impact of the information environment on operations continues to increase. What Army forces do to achieve advantages across it—information superiority—significantly affects the outcome of operations. Consequently, FM 3-0 revises how the Army views information operations and staff responsibility for associated Army information tasks. Other changes include replacing the battlefield operating systems with the warfighting functions and adding the spectrum of conflict with related operational themes.

Chaos, chance, and friction dominate land operations as much today as when Clausewitz wrote about them after the Napoleonic wars. In this environment, an offensive mindset—the predisposition to seize, retain, and exploit the initiative to positively change the situation—makes combat power decisive. The high quality of Army leaders and Soldiers is best exploited by allowing subordinates maximum latitude to exercise individual and small-unit initiative. Tough, realistic training prepares leaders for this, and FM 3-0 prescribes giving them the maximum latitude to accomplish the mission successfully. This requires a climate of trust in the abilities of superior and subordinate alike. It also requires leaders at every level to think and act flexibly, constantly adapting to the situation. Subordinates' actions are guided by the higher commander's intent but not circumscribed by excessive control. This is a continuing tension across the Army, aggravated by advanced information systems that can provide higher commanders with the details of lower echelon operations. The temptation for senior leaders to micromanage subordinates is great, but it must be resisted.

Despite the vital importance of nonlethal action to change the civil situation, FM 3-0 recognizes that the Army's primary purpose is deterrence, and should deterrence fail, decisively winning the Nation's wars by fighting within an interdependent joint team. America is at war and should expect to remain fully engaged for the next several decades in a persistent conflict against an enemy dedicated to U.S. defeat as a nation and eradication as a society. This conflict will be waged in an environment that is complex, multidimensional, and rooted in the human dimension. This conflict cannot be won by military forces alone; it requires close cooperation and coordination of diplomatic, informational, military, and economic efforts. Due to the human nature of the conflict, however, landpower will remain important to the military effort and essential to victory. FM 3-0 considers the nature of today's enemies as well as a wide range of other potential threats. It contains doctrine that seeks nothing less than victory for the United States—now and in the future.

As with all previous Army capstone doctrine, this doctrine provides direction for the Army and reflects its progress through the years. Like the manual that emerged from Valley Forge, it reflects the lessons learned from combat experience and addresses strategic, operational, and tactical realities. Baron von Steuben's doctrine allowed for the creation of forces capable of standing against the British Army, the world's best, by giving the Continental Army the skills necessary to win. Then, as now, success depended on the determination of well-trained Soldiers, the quality of their small-unit leadership, and the abilities of their commanders.

Chapter 1

The Operational Environment

Unfortunately, the dangers and challenges of old have been joined by new forces of instability and conflict.... A new and more malignant form of global terrorism rooted in extremist and violent jihadism; New manifestations of ethnic, tribal, and sectarian conflict all over the world; The proliferation of weapons of mass destruction; Failed and failing states; States enriched with oil profits and discontented with the current international order; and Centrifugal forces in other countries that threaten national unity, stability, and internal peace—but also with implications for regional and global security. Worldwide, there are authoritarian regimes facing increasingly restive populations seeking political freedom as well as a better standard of living. And finally, we see both emergent and resurgent great powers whose future path is still unclear.

Robert M. Gates
Secretary of Defense

Military operations occur within a complex framework of environmental factors that shape their nature and affect their outcomes. This requires a broad understanding of the strategic and operational environment and their relevance to each mission. This includes the characteristics of the particular operational environment to each mission and how aspects of the environment become essential elements in shaping how Army forces conduct operations. This chapter discusses the operational environment as the basis for understanding the Army's doctrine for the conduct of land operations. It addresses these operations, emphasizing the Army's expeditionary and campaign qualities and the integral role of Army forces in unified actions—joint, interagency, and multinational undertakings that execute campaigns and major operations.

INSTABILITY AND PERSISTENT CONFLICT

1-1. *Operational environments* are a composite of the conditions, circumstances, and influences that affect the employment of capabilities and bear on the decisions of the commander (JP 3-0). While they include all enemy, adversary, friendly, and neutral systems across the spectrum of conflict, they also include an understanding of the physical environment, the state of governance, technology, local resources, and the culture of the local population. This doctrine pertains in an era of complex local, regional, and global change leading to both opportunities and risks. This risk component of this change manifests in certain trends that drive instability and a continuing state of persistent conflict. Some important trends that will affect ground force operations in an era of persistent conflict include:

- Globalization.
- Technology.
- Demographic changes.
- Urbanization.
- Resource demand.
- Climate change and natural disasters.
- Proliferation of weapons of mass destruction and effects.
- Failed or failing states.

1-2.   Globalization will continue to affect global prosperity positively but is also believed to export terror worldwide. Interdependent economies have enabled great wealth. The benefits of this wealth remain concentrated in the hands of a few while the risks of failure are borne by many. This unequal distribution of wealth often creates "have" and "have-not" conditions that can spawn conflict. This dichotomy is evident between developed nations in the northern hemisphere and developing nations to their south and in the southern hemisphere. By 2015, experts project that up to 2.8 billion people—almost exclusively in economic "have-not" areas in developing nations—will live below the poverty level. These people are more vulnerable to recruitment by extremist groups. Globalization has also contributed to the rise of nonstate actors to economic, informational, and even military and diplomatic positions rivaling or exceeding those of states. The decline in state power and influence makes diplomatic interaction more difficult and complex. Globalization has already left several states behind, and more nations will lag in the increasing tempo of globalization. As a result, their populations will both suffer and become more apt to embrace radical ideologies to express their frustration and increase their desire, if not ability, to share in global prosperity.

1-3.   Technology will be another double-edged sword. Often, innovations that improve the quality of life and livelihood are also used by adversaries to destroy those lives. It would seem as though technology is an asymmetric advantage of developed nations. They have greater access to research facilities to develop and innovate. Technology also gives nations access to the industrial base. These nations can then mass-produce advanced products and widely distribute them at relatively low costs. The low cost of products, their user-friendly design, and their availability in a global economy makes advanced technology accessible to unstable states as well as extremist organizations. The revolution and proliferation of benefits derived from integrating multidisciplinary nano- and bio-technologies and smart materials potentially promises to improve living conditions. However, these products will not always be available at the pace and in the quantities necessary to make them and their benefits as universally available as desired. This disparity can create another source of friction between the haves and have-nots. Moreover, the proliferation, falling costs, and availability of technologically advanced products—especially expanded information technologies using mobile networks and expanded use of wireless and global fiber-optic networks—enable nonstate adversaries to acquire them.

1-4.   Population growth in the developing world will increase opportunities for instability, radicalism, and extremism. Populations of some less-developed countries will almost double by 2020, most notably in Africa, the Middle East, and South and Southeast Asia. The "youth bulge" created by this growth will be vulnerable to antigovernment and radical ideologies, worsening governance challenges. Middle class populations will grow as well. They will demand improved quality-of-life benefits and more resources to go with their increased wealth. The middle class population of India already exceeds the entire population of the United States. Inability or inequity to distribute wealth will intensify tensions between haves and have-nots. It will likely escalate calls for changes in how to share wealth globally.

1-5.   Urbanization will characterize well over half the world's population by 2015. By 2030, up to 60 percent will be urban-dwellers. Many cities are already huge; 15 have populations in excess of 10 million. Eight of the megacities lie near known geological fault lines that threaten natural disaster. These megacities increasingly assume the significance of nation-states, posing similar governance and security concerns. Their urban growth is much more pronounced in developing regions where states are already more prone to failure. Organized crime and extremist ideological and cultural enclaves will flourish in urban terrain, overwhelming and supplanting local governance apparatus. Chronic unemployment, over-crowding, pollution, uneven resource distribution, and poor sanitation, health, and other basic services will add to population dissatisfaction and increase the destructive allure of radical ideologies.

1-6.   Demand for energy, water, and food for growing populations will increase competition and, potentially, conflict. Resources—especially water, gas, and oil—are finite. By 2030, energy consumption will probably exceed production. Current sources, investment, and development of alternatives will not bridge the gap, according to the best estimates. A shift to cleaner fuels such as natural gas will find about 60 percent of known reserves concentrated in Russia, Iran, and Qatar. Demand for water doubles every 20 years. By 2015, 40 percent of the world's population will live in "water-stressed" countries, increasing the potential for competition over a resource that has already led to conflict in the past.

1-7.   Climate change and natural disasters will compound already difficult conditions in developing countries. They will cause humanitarian crises, driving regionally destabilizing population migrations and raising potential for epidemic diseases. Desertification is occurring at nearly 50–70 thousand square miles per year. Over 15 million people die each year from communicable diseases; these numbers may grow exponentially as urban densities increase. Increased consumption of resources, especially in densely populated areas, will increase air, water, land, and potentially even space pollution. Depletion of resources will also compound this problem. Depletion reduces natural replenishment sources as well as intensifies the effects of natural disasters, having increasingly greater impacts on more densely populated areas.

1-8.   Proliferation of weapons of mass destruction and effects will increase the potential for catastrophic attacks. These attacks will be destabilizing globally and undercut the confidence that spurs economic development. The threat of the use of weapons of mass destruction is as real as it is deadly. Over 1,100 identified terrorist organizations exist. Some of them, most notably Al Qaeda, actively seek weapons of mass destruction. In nuclear proliferation, there were 662 reported incidents of unauthorized activities surrounding nuclear and radioactive materials since 1993. These involved quantities of enriched uranium from military and civilian reactors in excess of 3,700 tons, enough to produce thousands of nuclear weapons. Additionally, some nuclear nations are sharing technology as a means to earn money and secure influence. For small countries and terrorist organizations, biological weapons convey a similar status as nuclear weapons. Many biological and chemical agents are produced easily and cheaply. Wider Internet access has made the technologies and processes of developing weapons of mass destruction and effects readily available to potential adversaries. Further, some states may pursue these programs to assure their security and prevent forced regime change.

1-9.   Governments of nation-states are facing increasingly greater challenges in providing effective support to their growing populations. Security, economic prosperity, basic services, and access to resources strain systems designed in an industrial age. Additionally, these governments are unprepared to increase openness intellectually or culturally to deal with an information age. Compounding this inability to adapt, state governments find themselves pitted against those that have made the shift and are already exploiting it to gain support of local populaces. These adversaries can include criminal organizations, extremist networks, private corporate enterprises, and increasingly powerful megacities. Stability will be paramount, not the form of governance. The problem of failed or failing states can result in the formation of safe havens in which adversaries can thrive.

INFLUENCES ON THE OPERATIONAL ENVIRONMENT

1-10. The driving forces, trends, and variables discussed above create a solid framework upon which to build a picture of the conditions that will shape an era of persistent conflict. Science and technology, information technology, transportation technology, the acceleration of the global economic community, and the rise of a networked society will all impact the operational environment. The international nature of commercial and academic efforts will also have dramatic impacts. The complexity of the operational environment will guarantee that future operations will occur across the spectrum of conflict.

1-11. In essence, the operational environment of the future will still be an arena in which bloodshed is the immediate result of hostilities between antagonists. It will also be an arena in which operational goals are attained or lost not only by the use of highly lethal force but also by how quickly a state of stability can be established and maintained. The operational environment will remain a dirty, frightening, physically and emotionally draining one in which death and destruction result from environmental conditions creating humanitarian crisis as well as conflict itself. Due to the extremely high lethality and range of advanced weapons systems, and the tendency of adversaries to operate among the population, the risk to combatants and noncombatants will be much greater. All adversaries, state or nonstate, regardless of technological or military capability, can be expected to use the full range of options, including every political, economic, informational, and military measure at their disposal. In addition, the operational environment will expand to areas historically immune to battle, including the continental United States and the territory of multinational partners, especially urban areas. In fact, the operational environment will probably include areas not defined by geography, such as cyberspace. Computer network attacks will span borders and will be able to hit anywhere, anytime. With the exception of cyberspace, all operations will be conducted "among the people" and outcomes will be measured in terms of effects on populations.

1-12. The operational environment will be extremely fluid, with continually changing coalitions, alliances, partnerships, and actors. Interagency and joint operations will be required to deal with this wide and intricate range of players occupying the environment. International news organizations, using new information and communications technologies, will no longer have to depend on states to gain access to the area of operations and will greatly influence how operations are viewed. They will have satellites or their own unmanned aerial reconnaissance platforms from which to monitor the scene. Secrecy will be difficult to maintain, making operations security more vital than ever. Finally, complex cultural, demographic, and physical environmental factors will be present, adding to the fog of war. Such factors include humanitarian crises, ethnic and religious differences, and complex and urban terrain, which often become major centers of gravity and a haven for potential threats. The operational environment will be interconnected, dynamic, and extremely volatile.

THE CHANGING NATURE OF THE THREAT

1-13. States, nations, transnational actors, and nonstate entities will continue to challenge and redefine the global distribution of power, the concept of sovereignty, and the nature of warfare. Threats are nation-states, organizations, people, groups, conditions, or natural phenomena able to damage or destroy life, vital resources, or institutions. Preparing for and managing these threats requires employing all instruments of national power—diplomatic, informational, military, and economic. Threats may be described through a range of four major categories or challenges: traditional, irregular, catastrophic, and disruptive. While helpful in describing the threats the Army is likely to face, these categories do not define the nature of the adversary. In fact, adversaries may use any and all of these challenges in combination to achieve the desired effect against the United States.

1-14. Traditional threats emerge from states employing recognized military capabilities and forces in understood forms of military competition and conflict. In the past, the United States optimized its forces for this challenge. The United States currently possesses the world's preeminent conventional and nuclear forces, but this status is not guaranteed. Many nations maintain powerful conventional forces, and not all are friendly to the United States. Some of these potentially hostile powers possess weapons of mass destruction. Although these powers may not actively seek armed confrontation and will actively avoid U.S. military strength, their activities can provoke regional conflicts that threaten U.S. interests. Deterrence therefore remains the first aim of the joint force. Should deterrence fail, and there is some evidence that deterrence is less able to accomplish this goal, the United States strives to maintain capabilities to overmatch any combination of enemy conventional and unconventional forces.

1-15. Irregular threats are those posed by an opponent employing unconventional, asymmetric methods and means to counter traditional U.S. advantages. A weaker enemy often uses irregular warfare to exhaust the U.S. collective will through protracted conflict. Irregular warfare includes such means as terrorism, insurgency, and guerrilla warfare. Economic, political, informational, and cultural initiatives usually accompany and may even be the chief means of irregular attacks on U.S. influence.

1-16. Catastrophic threats involve the acquisition, possession, and use of nuclear, biological, chemical, and radiological weapons, also called weapons of mass destruction and effects. Possession of these weapons gives an enemy the potential to inflict sudden and catastrophic effects. The proliferation of related technology has made this threat more likely than in the past.

1-17. Disruptive threats involve an enemy using new technologies that reduce U.S. advantages in key operational domains. Disruptive threats involve developing and using breakthrough technologies to negate current U.S. advantages in key operational domains.

1-18. By combining traditional, disruptive, catastrophic, and irregular capabilities, adversaries will seek to create advantageous conditions by quickly changing the nature of the conflict and moving to employ capabilities for which the United States is least prepared. The enemy will seek to interdict U.S. forces attempting to enter any area of crisis. If U.S. forces successfully gain entry, the enemy will seek engagement in complex terrain and urban environments as a way of offsetting U.S. advantages. Methods used by adversaries include dispersing their forces into small mobile combat teams—combined only when required to strike a common objective—and becoming invisible by blending in with the local population.

1-19. Threats can be expected to use the environment and rapidly adapt. Extremist organizations will seek to take on state-like qualities using the media and technology and their position within a state's political, military, and social infrastructures to their advantage. Their operations will become more sophisticated, combining conventional, unconventional, irregular, and criminal tactics. They will focus on creating conditions of instability, seek to alienate legitimate forces from the population, and employ global networks to expand local operations. The threat will employ advanced information operations and will not be bound by limits on the use of violence.

1-20. Future conflicts are much more likely to be fought "among the people" instead of "around the people." This fundamentally alters the manner in which Soldiers can apply force to achieve success in a conflict. Enemies will increasingly seek populations within which to hide as protection against the proven attack and detection means of U.S. forces, in preparation for attacks against communities, as refuge from U.S. strikes against their bases, and to draw resources. War remains a battle of wills—a contest for dominance over people. The essential struggle of the future conflict will take place in areas in which people are concentrated and will require U.S. security dominance to extend across the population.

OPERATIONAL AND MISSION VARIABLES

1-21. The operational environment includes physical areas—the air, land, maritime, and space domains. It also includes the information that shapes the operational environment as well as enemy, adversary, friendly, and neutral systems relevant to that joint operation. The operational environment for each campaign or major operation is different, and it evolves as each campaign or operation progresses. Army forces use operational variables to understand and analyze the broad environment in which they are conducting operations. They use mission variables to focus analysis on specific elements of the environment that apply to their mission.

OPERATIONAL VARIABLES

1-22. Military planners describe the operational environment in terms of operational variables. Operational variables are those broad aspects of the environment, both military and nonmilitary, that may differ from one operational area to another and affect campaigns and major operations. Operational variables describe not only the military aspects of an operational environment but also the population's influence on it. Joint planners analyze the operational environment in terms of six interrelated operational variables: political, military, economic, social, information, and infrastructure. To these variables Army doctrine adds two more: physical environment and time. As a set, these operational variables are often abbreviated as PMESII-PT.

1-23. The variables provide a view of the operational environment that emphasizes its human aspects. Since land forces operate among populations, understanding the human variables is crucial. They help describe each operation's context for commanders and other leaders. Understanding them helps commanders appreciate how the military instrument complements the other instruments of national power. Comprehensive analysis of the variables usually occurs at the joint level; Army commanders continue analysis to improve their understanding of their environment. The utility of the operational variables improves with flexible application; human societies are very complicated and defy precise "binning." Whenever possible, commanders and staffs utilize specialists in each variable to improve analysis.

Political

1-24. The political variable describes the distribution of responsibility and power at all levels of governance. Political structures and processes enjoy varying degrees of legitimacy with populations from the local through international levels. Formally constituted authorities and informal or covert political powers strongly influence events. Political leaders can use ideas, beliefs, actions, and violence to enhance their power and control over people, territory, and resources. Many sources of political motivation exist. These may include charismatic leadership; indigenous security institutions; and religious, ethnic, or economic communities. Political opposition groups or parties also affect the situation. Each may deal differently with U.S. or multinational forces. Understanding the political circumstances helps commanders and staffs recognize key organizations and determine their aims and capabilities.

1-25. Understanding political implications requires analyzing all relevant partnerships—political, economic, military, religious, and cultural. This analysis captures the presence and significance of external organizations and other groups. These include groups united by a common cause. Examples are private security organizations, transnational corporations, and nongovernmental organizations that provide humanitarian assistance.

1-26. A political analysis also addresses the effect of will. Will is the primary intangible factor; it motivates participants to sacrifice to persevere against obstacles. Understanding the motivations of key groups (for example, political, military, and insurgent) helps clarify their goals and willingness to sacrifice to achieve their ends.

1-27. The political variable includes the U.S. domestic political environment. Therefore, mission analysis and monitoring the situation includes an awareness of national policy and strategy.

Military

1-28. The military variable includes the military capabilities of all armed forces in a given operational environment. For many states, an army is the military force primarily responsible for maintaining internal and external security. Paramilitary organizations and guerrilla forces may influence friendly and hostile military forces. Militaries of other states not directly involved in a conflict may also affect them. Therefore, analysis should include the relationship of regional land forces to the other variables. Military analysis examines the capabilities of enemy, adversary, host-nation, and multinational military organizations. Such capabilities include—

- Equipment.
- Manpower.
- Doctrine.
- Training levels.
- Resource constraints.
- Leadership.
- Organizational culture.
- History.
- Nature of civil-military relations.

Understanding these factors helps commanders estimate the actual capabilities of each armed force. Analysis should focus on each organization's ability to field capabilities and use them domestically, regionally, and globally.

Economic

1-29. The economic variable encompasses individual and group behaviors related to producing, distributing, and consuming resources. Specific factors may include the influence of—

- Industrial organizations.
- Trade.
- Development (including foreign aid).
- Finance.
- Monetary policy and conditions.
- Institutional capabilities.

- Geography.
- Legal constraints (or the lack of them) on the economy.

1-30. While the world economy is becoming interdependent, local economies differ. These differences significantly influence political choices, including individuals' decisions to support or subvert the existing order. Many factors create incentives or disincentives for individuals and groups to change the economic status quo. These may include—

- Technical knowledge.
- Decentralized capital flows.
- Investment.
- Price fluctuations.
- Debt.
- Financial instruments.
- Protection of property rights.
- Existence of black market or underground economies.

Thus, indicators measuring potential benefits or costs of changing the political-economic order may enhance understanding the social and behavioral dynamics of friendly, adversary, and neutral entities.

Social

1-31. The social variable describes societies within an operational environment. A society is a population whose members are subject to the same political authority, occupy a common territory, have a common culture, and share a sense of identity. Societies are not monolithic. They include diverse social structures. Social structure refers to the relations among groups of persons within a system of groups. It includes institutions, organizations, networks, and similar groups. (FM 3-24 discusses socio-cultural factors analysis and social network analysis.)

1-32. Culture comprises shared beliefs, values, customs, behaviors, and artifacts that society members use to cope with their world and with one another. Societies usually have a dominant culture but may have many secondary cultures. Different societies may share similar cultures, but societal attributes change over time. Changes may occur in any of the following areas:

- Demographics.
- Religion.
- Migration trends.
- Urbanization.
- Standards of living.
- Literacy and nature of education.
- Cohesiveness and activity of cultural, religious, or ethnic groups.

Social networks, social status and related norms, and roles that support and enable individuals and leaders require analysis. This analysis should also address societies outside the operational area whose actions, opinions, or political influence can affect the mission.

1-33. People base their actions on perceptions, assumptions, customs, and values. Cultural awareness helps identify points of friction within populations, helps build rapport, and reduces misunderstandings. It can improve a force's ability to accomplish its mission and provide insight into individual and group intentions. However, cultural awareness requires training before deploying to an unfamiliar operational environment and continuous updating while deployed. Commanders develop their knowledge of the societal aspects within their areas of operations to a higher level of cultural astuteness, one that allows them to understand the impact of their operations on the population and prepares them to meet local leaders face-to-face.

Information

1-34. Joint doctrine defines the *information environment* as the aggregate of individuals, organizations, and systems that collect, process, disseminate, or act on information (JP 3-13). The environment shaped by information includes leaders, decisionmakers, individuals, and organizations. The global community's access and use of data, media, and knowledge systems occurs in the information shaped by the operational environment. Commanders use information engagement to shape the operational environment as part of their operations. (Paragraphs 7-10 through 7-22 discuss information engagement.)

1-35. Media representatives significantly influence the information that shapes the operational environment. Broadcast and Internet media sources can rapidly disseminate competing views of military operations worldwide. Adversaries often seek to further their aims by controlling and manipulating how audiences at all levels perceive a situation's content and context. Media coverage influences U.S. political decisionmaking, popular opinion, and multinational sensitivities.

1-36. Complex telecommunications networks now provide much of the globe with a vast web of communications capabilities. Observers and adversaries have unprecedented access to multiple information sources. They often attempt to influence opinion by providing their own interpretation of events. Televised news and propaganda reach many people. However, in developing countries, information still may flow by less sophisticated means such as messengers and graffiti. Understanding the various means of communications is important. Observers and adversaries control information flow and influence audiences at all levels.

Infrastructure

1-37. Infrastructure comprises the basic facilities, services, and installations needed for a society's functioning. Degrading infrastructure affects the entire operational environment. Infrastructure also includes technological sophistication—the ability to conduct research and development and apply the results to civil and military purposes.

1-38. Not all segments of society view infrastructure in the same way. Improvements viewed by some as beneficial may not be perceived as such by all. One community may perceive certain improvements as favoring other communities at its expense. Effective information engagement is necessary to address such concerns. Actions affecting infrastructure require a thorough analysis of possible effects.

Physical Environment

1-39. The physical environment includes the geography and man-made structures in the operational area. The following factors affect the physical environment:
- Man-made structures, particularly urban areas.
- Climate and weather.
- Topography.
- Hydrology.
- Natural resources.
- Biological features and hazards.
- Other environmental conditions.

The enemy understands that less complex and open terrain often exposes their military weaknesses. Therefore, they may try to counteract U.S. military advantages by operating in urban or other complex terrain and during adverse weather conditions.

Time

1-40. Time is a significant consideration in military operations. Analyzing it as an operational variable focuses on how an operation's duration might help or hinder each side. This has implications at every planning level. An enemy with limited military capability usually views protracted conflict as advantageous to them. They avoid battles and only engage when conditions are overwhelmingly in their favor. This is a strategy of exhaustion. Such a strategy dominated the American Revolution and remains effective today.

The enemy concentrates on surviving and inflicting friendly and civilian casualties over time. Although the military balance may not change, this creates opportunities to affect the way domestic and international audiences view the conflict. Conversely, a hostile power may attempt to mass effects and achieve decisive results in a short period.

MISSION VARIABLES

1-41. The operational variables are directly relevant to campaign planning; however, they may be too broad for tactical planning. That does not mean that they are not valuable at the tactical level; they are fundamental to developing an understanding of the operational environment necessary to plan at any level, in any situation. The degree to which each operational variable provides useful information depends on the situation and echelon. For example, social and economic variables often receive close analysis as part of enemy and civil considerations at brigade and higher levels. They may affect the training and preparation of small units. However, they may not be relevant to a small-unit leader's mission analysis. That leader may only be concerned with such questions as "Who is the tribal leader for this village?" "Is the electrical generator working?" "Does the enemy have antitank missiles?"

1-42. Upon receipt of a warning order or mission, Army tactical leaders narrow their focus to six mission variables. Mission variables are those aspects of the operational environment that directly affect a mission. They outline the situation as it applies a specific Army unit. The mission variables are mission, enemy, terrain and weather, troops and support available, time available and civil considerations (METT-TC). These are the categories of relevant information used for mission analysis. Army leaders use the mission variables to synthesize operational variables and tactical-level information with local knowledge about conditions relevant to their mission. (Paragraphs 5-24 through 5-39 expand the discussion of the mission variables.)

1-43. Army forces interact with people at many levels. In general, the people in any operational area can be categorized as enemies, adversaries, supporters, and neutrals. One reason land operations are complex is that all four categories are intermixed, often with no easy means to distinguish one from another. They are defined as—

- An *enemy* **is a party identified as hostile against which the use of force is authorized.** An enemy is also called a combatant and is treated as such under the law of war.
- An *adversary* is a party acknowledged as potentially hostile to a friendly party and against which the use of force may be envisaged (JP 3-0). Adversaries include members of the local populace who sympathize with the enemy.
- A *supporter* **is a party who sympathizes with friendly forces and who may or may not provide material assistance to them.**
- A *neutral* **is a party identified as neither supporting nor opposing friendly or enemy forces.**

1-44. Incorporating the analysis of the operational variables into METT-TC emphasizes the operational environment's human aspects. This emphasis is most obvious in civil considerations, but it affects the other METT-TC variables as well. Incorporating human factors into mission analysis requires critical thinking, collaboration, continuous learning, and adaptation. It also requires analyzing local and regional perceptions. Many factors influence perceptions of the enemy, adversaries, supporters, and neutrals. These include—

- Language.
- Culture.
- Geography.
- History.
- Education.
- Beliefs.
- Perceived objectives and motivation.
- Communications media.
- Personal experience.

UNIFIED ACTION

1-45. *Unified action* is the synchronization, coordination, and/or integration of the activities of governmental and nongovernmental entities with military operations to achieve unity of effort (JP 1). It involves the application of all instruments of national power, including actions of other government agencies and multinational military and nonmilitary organizations. Combatant commanders play a pivotal role in unifying actions; however, subordinate commanders also integrate and synchronize their operations directly with the activities and operations of other military forces and nonmilitary organizations in their area of operations. Department of Defense and other government agencies often refer to the unified action environment as joint, interagency, intergovernmental, and multinational.

1-46. Unified action includes joint integration. Joint integration extends the principle of combined arms to operations conducted by two or more Service components. The combination of diverse joint force capabilities generates combat power more potent than the sum of its parts. Joint integration does not require joint command at all echelons. It does, however, require joint interoperability at all levels. Army mission accomplishment links to the national strategic end state through joint campaigns and major operations.

CAMPAIGNS AND JOINT OPERATIONS

1-47. Joint planning integrates military power with other instruments of national power to achieve the desired military end state. (The *end state* is the set of required conditions that defines achievement of the commander's objectives [JP 3-0].) This planning connects the strategic end state to campaign design and ultimately to tactical missions. Joint force commanders use campaigns and joint operations to translate their operational-level actions into strategic results. A *campaign* is a series of related major operations aimed at achieving strategic and operational objectives within a given time and space (JP 5-0). Campaigns are always joint operations.

1-48. Campaigns exploit the advantages of interdependent Service capabilities through unified action. Coordinated, synchronized, and integrated action is necessary to reestablish civil authority after joint operations end, even when combat is not required. Effective joint and Army operations require all echelons to perform extensive collaborative planning and understand joint interdependence.

JOINT INTERDEPENDENCE

1-49. Joint interdependence is the purposeful reliance by one Service's forces on another Service's capabilities to maximize the complementary and reinforcing effects of both. Army forces operate as part of an interdependent joint force. Joint capabilities make Army forces more effective than they would be otherwise. Combinations of joint capabilities defeat enemy forces by shattering their ability to operate as a coherent, effective whole. Acting with other instruments of national power, joint forces also work to reduce the level of violence and establish security. (Table 1-1 lists areas of joint interdependence that directly enhance Army operations.)

Table 1-1. Areas of joint interdependence

Joint command and control. Integrated capabilities that—
- Gain information superiority through improved, fully synchronized, integrated intelligence, surveillance, and reconnaissance; knowledge management; and information management.
- Share a common operational picture.
- Improve the ability of joint force and Service component commanders to conduct operations.

Joint intelligence. Integrated processes that—
- Reduce unnecessary redundancies in collection asset tasking through integrated intelligence, surveillance, and reconnaissance.
- Increase processing and analytic capability.
- Facilitate collaborative analysis.
- Provide global intelligence production and dissemination.
- Provide intelligence products that enhance situational understanding by describing and assessing the operational environment.

Joint information operations capabilities. Integrated capabilities, including—
- Special technical operations.
- Electronic warfare platforms and personnel.
- Reachback to strategic assets.

Joint fires. Integrated fire control networks that allow joint forces to deliver coordinated fires from two or more Service components.

Joint air operations. Air and Naval forces able to—
- Maneuver aircraft to positions of advantage over the enemy beyond the reach of land forces.
- Gain and maintain air superiority that extends the joint force's area of influence by providing freedom from attack as well as freedom to attack.
- Support operational and tactical maneuver with lethal and nonlethal fires.

Joint air and missile defense. A comprehensive joint protection umbrella that—
- Begins with security of ports of debarkation.
- Enables uninterrupted force flow against diverse anti-access threats.
- Extends air and missile defense to multinational partners.

Joint force projection. Strategic and operational lift capabilities and automated planning processes that facilitate strategic responsiveness and operational agility.

Joint sustainment. Deliberate, mutual reliance by each Service component on the sustainment capabilities of two or more Service components. It can reduce redundancies or increase the robustness of operations without sacrificing effectiveness.

Joint space operations. Access to national imagery, communications, satellite, and navigation capabilities that enhance situational awareness and support understanding of the operational environment.

1-50. The other Services rely on Army forces to complement their capabilities. (Table 1-2, page 1-12, lists Army capabilities that enhance other Service component operations.)

Table 1-2. Army capabilities that complement other Services

Security and control of terrain, people, and resources including—
- Governance over an area or region.
- Protection of key infrastructure and facilities from ground threats.

Land-based ballistic missile defense, including defense against cruise missiles and counterrocket, counterartillery, and countermortar capabilities.

Chemical, biological, radiological, and nuclear operations.

Support to interagency reconstruction efforts and provision of essential services to an affected population.

Denial of sanctuary through ground maneuver, enabling attack from the air.

Discriminate force application within populated areas.

Inland sustainment of bases and of forces operating from those bases.

Land operations against enemy air and sea bases.

Detainee and enemy prisoner of war operations.

Intelligence support.

1-51. Joint forces also rely on Army forces for support and services as designated in—
- Title 10, U.S. Code.
- Other applicable U.S. laws.
- Department of Defense implementation directives and instructions.
- Inter-Service agreements.
- Multinational agreements.
- Other applicable authorities and Federal regulations.

This support and other support directed by combatant commanders are broadly defined as "Army support to other Services."

INTERAGENCY COORDINATION AND COOPERATION WITH OTHER ORGANIZATIONS

1-52. Interagency coordination is inherent in unified action. Within the context of Department of Defense involvement, *interagency coordination* is the coordination that occurs between elements of Department of Defense and engaged U.S. Government agencies for the purpose of achieving an objective (JP 3-0). In addition, unified action involves synchronizing joint or multinational military operations with activities of other government agencies, intergovernmental organizations, nongovernmental organizations, and contractors. During civil support operations, unified action includes local and state government agencies. It occurs at every level—tactical, operational, and strategic.

Civilian Organizations

1-53. Commanders must understand the respective roles and capabilities of civilian organizations in unified action. Other agencies of the Federal government work with the military and are part of a national chain of command under the President of the United States. While this does not guarantee seamless integration, it does provide a legal basis for cooperation. Although experience and professional culture differ widely, each organization needs to recognize and capitalize on the inherent professionalism of the other to develop the teamwork necessary for the campaign.

1-54. Most civilian organizations are not under military control. Nor does the U.S. ambassador or a United Nations commissioner control them. Civilian organizations have different organizational cultures and norms. Some may be willing to work with Army forces; others may not. Thus, personal contact and trust building are essential. Command emphasis on immediate and continuous coordination encourages effective cooperation. Commanders should establish liaison with civilian organizations to integrate their efforts as much as possible with Army and joint operations. Civil affairs units typically establish this liaison.

1-55. Civilian organizations bring resources and capabilities that can help establish host-nation civil authority and capabilities. However, civilian agencies may arrive well after military operations have begun. Therefore, joint and Army forces prepare to establish and maintain order if host-nation authorities cannot do so. Successfully performing these tasks can help secure a lasting peace and facilitate the timely withdrawal of U.S. and multinational forces.

1-56. Army forces provide sustainment and security for civilian organizations when directed, since many of these organizations lack these capabilities. Army forces often provide this support to state and local agencies during civil support operations. (Table 1-3 lists examples of civilian organizations.)

Table 1-3. Definitions and examples of civilian organizations

Category	Definition	Examples
Other government agency	Within the context of interagency coordination, a non-Department of Defense agency of the United States Government (JP 1).	• Department of State • Central Intelligence Agency • Federal Bureau of Investigation • National Security Agency • U.S. Agency for International Development
Intergovernmental organization	An organization created by a formal agreement (for example, a treaty) between two or more governments. It may be established on a global, regional, or functional basis for wide-ranging or narrowly defined purposes. Formed to protect and promote national interests shared by member states (JP 3-08).	• United Nations • European Union • North Atlantic Treaty Organization • Organization for Security and Cooperation in Europe • African Union
Nongovernmental organization	A private, self-governing, not-for-profit organization dedicated to alleviating human suffering; and/or promoting education, health care, economic development, environmental protection, human rights, and conflict resolution; and/or encouraging the establishment of democratic institutions and civil society (JP 3-08).	See the United Nations Web site (www.un.org) to research accredited nongovernmental organizations.

Contractors

1-57. A *contractor* is a person or business that provides products or services for monetary compensation. A contractor furnishes supplies and services or performs work at a certain price or rate based on the terms of a contract (FM 3-100.21). Contracted support often includes traditional goods and services support but may include interpreter communications, infrastructure, and other related support. In military operations, contractors may provide—

- Life support.
- Construction and engineering support.
- Weapons systems support.
- Security.
- Other technical services.

(FM 3-100.21 contains doctrine for contractors accompanying deployed forces.)

MULTINATIONAL OPERATIONS

1-58. *Multinational operations* is a collective term to describe military actions conducted by forces of two or more nations, usually undertaken within the structure of a coalition or alliance (JP 3-16). In multinational operations, all parties agree to the commitment of forces, even if the resources each invests are different. While each nation has its own interests, all nations bring value to the operation. Each national force has unique capabilities, and each usually contributes to the operation's legitimacy in terms of international or local acceptability.

1-59. An *alliance* is the relationship that results from a formal agreement (for example, a treaty) between two or more nations for broad, long-term objectives that further the common interests of the members (JP 3-0). Military alliances, such as the North Atlantic Treaty Organization (NATO), allow partners to establish formal, standard agreements. For example, U.S. forces operate within a highly developed multinational command structure to maintain the armistice on the Korean peninsula. Alliance members strive for interoperability. They field compatible military systems, establish common procedures, and develop contingency plans to meet potential threats.

1-60. A *coalition* is an ad hoc arrangement between two or more nations for common action (JP 5-0). Nations usually form coalitions for focused, short-term purposes. A *coalition action* is a multinational action outside the bounds of established alliances, usually for single occasions or longer cooperation in a narrow sector of common interest (JP 5-0). Coalition actions may be conducted under the authority of a United Nations resolution. Since coalition actions are not structured around formal treaties, a preliminary understanding of the requirements for operating with a specific foreign military may occur through peacetime military engagement. (Paragraph 2-17 defines peacetime military engagement.)

1-61. Agreement among the multinational partners establishes the level of command authority vested in a multinational force commander. The President retains command authority over U.S. forces. Most nations have similar restrictions. However, in certain circumstances, it may be prudent or advantageous to place Army forces under the operational control of a multinational commander. Often, multinational organizations have complex lines of command. To compensate for limited unity of command, multinational partners concentrate on achieving unity of effort. Consensus building, rather than direct command authority, is often the key element of successful multinational operations.

1-62. An Army officer assigned to command a multinational force faces many complex demands. These include dealing with cultural issues, different languages, interoperability challenges, national caveats on the use respective forces, and sometimes underdeveloped command and control. Multinational force commanders are also required to address different national procedures, restrictions, intelligence sharing, and theater sustainment functions. Another command challenge is the multinational commander's limited ability to choose or replace subordinates. Nations assign their contingent leaders. They answer to their national chains of command as well as to the multinational force commander. Every multinational operation differs. Commanders analyze the mission's peculiar requirements to exploit the multinational force's advantages and compensate for its limitations. (FM 6-22 discusses leadership considerations for multinational operations.)

1-63. Multinational sustainment requires detailed planning and coordination. Each nation normally provides a national support element to sustain its deployed forces. However, integrated multinational sustainment may improve efficiency and effectiveness. When directed, an Army theater sustainment command can provide logistic and other support to multinational forces. Integrating the support requirements of several national forces, often spread over considerable distances and across international boundaries, is challenging. Nonetheless, multinational partners can provide additional resources to address the sustainment challenges. For example, a multinational partner may provide a secure intermediate staging base near the operational area. Deploying and employing forces from an intermediate staging base is preferable to making a forcible entry from a distant base. This is especially true when the staging base offers a mature infrastructure.

1-64. During multinational operations, U.S. forces establish liaison with assigned multinational forces as soon as possible. Army forces exchange specialized liaison personnel based on mission requirements. Fields requiring specialized liaison may include aviation, fire support, engineer, intelligence, and civil affairs. Exchanging liaison fosters common understanding of missions and tactics, facilitates transfer of information, and enhances mutual trust and confidence.

1-65. Missions assigned to multinational units should reflect the capabilities and limitations of each national contingent. Some significant factors include—

- Relative size and mobility.
- Intelligence collection assets.
- Long-range fires capabilities.
- Special operations forces capabilities.
- Organic sustainment capabilities.
- Ability to contribute to theater air and missile defense.
- Training for operations in special environments.
- Willingness and ability to cooperate directly with troops of other nationalities.
- Preparation for defensive operations involving weapons of mass destruction.

1-66. When assigning missions, commanders should also consider special skills, language, and rapport with the local population as well as multinational partners' national sensitivities. Multinational commanders may assign host-nation forces home defense or police missions, such as sustainment area and base security. They may also entrust air defense, coastal defense, or a special operation to a single member of the multinational force based on that force's capabilities. Commanders should consider multinational force capabilities, such as mine clearance, that may exceed U.S. capabilities. (JP 3-16 and FM 100-8 contain doctrine for multinational operations. When revised, FM 100-8 will be renumbered FM 3-16.)

1-67. Since persistent conflict will affect a diverse range of international interests, it requires coalitions of nations joined in common cause to defeat a universal foe. Leaders at all levels must possess the ability to interoperate within a coalition requiring patience, understanding, and a willingness to subordinate self to the common good. Each nation will bring unique capabilities, strengths, and limitations to future coalitions, and Army forces must be able to operate within them. These coalitions will not always be composed of the same partners, and even when they are, the relative commitments of the partners will vary—sometimes even over time within the same conflict.

THE NATURE OF LAND OPERATIONS

War is thus an act of force to compel our enemy to do our will.

Carl von Clausewitz
On War[1]

1-68. Modern conflict occurs in many domains; however, landpower normally solidifies the outcome, even when it is not the decisive instrument. **Landpower is the ability—by threat, force, or occupation—to gain, sustain, and exploit control over land, resources, and people**. Landpower includes the ability to—

- Impose the Nation's will on an enemy, by force if necessary.
- Establish and maintain a stable environment that sets the conditions for political and economic development.
- Address the consequences of catastrophic events—both natural and man-made—to restore infrastructure and reestablish basic civil services.
- Support and provide a base from which joint forces can influence and dominate the air and maritime domains of an operational environment.

1-69. Several attributes of the land environment affect the application of landpower. These include—

- The requirement to deploy and employ Army forces rapidly.

---

[1] © 1984. Reproduced with permission of Princeton University Press.

- The requirement for Army forces to operate for protracted periods.
- The nature of close combat.
- Uncertainty, chance, friction, and complexity.

ARMY FORCES—EXPEDITIONARY AND CAMPAIGN CAPABILITIES

1-70. The initial operations in Afghanistan and Iraq were models of rapid, effective combat operations. Large enemy forces were destroyed or dispersed rapidly with little friendly loss. However, these operations also demonstrate that neither the duration nor the character of military campaigns is readily predictable. Future conflicts will include incomplete planning information, rapid deployments with little or no notice, and sustained operations in austere theaters. Joint, expeditionary warfare focuses on achieving decisive effects. It places a premium on promptly deploying landpower and constantly adapting to each campaign's unique circumstances as they change. But swift campaigns, however desirable, are the exception. Whenever objectives involve controlling populations or dominating terrain, campaign success usually requires employing landpower for protracted periods. Therefore, the Army combines expeditionary and campaign qualities to contribute decisive, sustained landpower to unified actions.

1-71. Expeditionary capability is the ability to promptly deploy combined arms forces worldwide into any operational environment and operate effectively upon arrival. Expeditionary operations require the ability to deploy quickly with little notice, shape conditions in the operational area, and operate immediately on arrival. Uncertainty as to the operational area, the possibility of a very austere environment, and the need to match forces to available lift drive expeditionary capabilities.

1-72. Expeditionary capabilities assure friends, allies, and foes that the Nation is able and willing to deploy the right combination of Army forces to the right place at the right time. Forward deployed units, forward positioned capabilities, peacetime military engagement, and force projection—from anywhere in the world—all contribute to expeditionary capabilities. Expeditionary capabilities enable the Army to respond rapidly under conditions of uncertainty to areas with complex and austere operational environments with the ability to fight not only on arrival but also through successive operations. Fast deploying and expansible Army forces provide the means to introduce operationally significant land forces into a crisis on short notice, providing preemptive options to deter, shape, fight and win if deterrence fails, and to sustain these options for the duration necessary to achieve success. Providing joint force commanders with expeditionary capability requires forces organized and equipped to be modular, versatile, and rapidly deployable with agile institutions capable of supporting them. Rapidly deployed expeditionary force packages provide immediate options for seizing or retaining the operational initiative. With their modular capabilities, these forces can be swiftly deployed, employed, and sustained for extended operations without an unwieldy footprint. These forces are tailored for the initial phase of operations, easily task-organized, and highly self-sufficient. Army installations worldwide serve as support platforms for force projection, providing capabilities and information on demand.

1-73. Expeditionary capabilities are more than physical attributes; they begin with a mindset that pervades the force. Soldiers with an expeditionary mindset are ready to deploy on short notice. They are confident that they can accomplish any mission. They are mentally and physically prepared to deploy anywhere in the world at any time in any environment against any adversary. Leaders with an expeditionary mindset are adaptive. They possess the individual initiative needed to accomplish missions through improvisation and collaboration. They are mentally prepared to operate within different cultures in any environment. An expeditionary mindset requires developing and empowering adaptive thinkers at all levels, from tactical to strategic.

1-74. Campaign capability is the ability to sustain operations as long as necessary to conclude operations successfully. Many conflicts are resolved only by altering the conditions that prompted the conflict. This requires combat power and time. The Army's campaign capability extends its expeditionary capability well beyond deploying combined arms forces that are effective upon arrival. It is an ability to conduct sustained operations for as long as necessary, adapting to unpredictable and often profound changes in the operational environment as the campaign unfolds. Army forces are organized, trained, and equipped for endurance. Their endurance stems from the ability to generate, protect, and sustain landpower—regardless of how far away it is deployed, how austere the environment, or how long the combatant commander requires it. It includes taking care not only of Soldiers but also of families throughout the complete cycle of deployment, employment, and redeployment. This involves anticipating requirements across the entire Army and making the most effective use of all available resources—deployed or not. Finally, campaign capability draws on iterative and continuous learning based on operational experience. This requirement extends to training at all echelons, from individual Soldier skills to operational-level collective tasks.

1-75. Campaigning requires a mindset and vision that complements expeditionary requirements. Soldiers understand that no matter how long they are deployed, the Army will take care of them and their families. They are confident that the loyalty they pledge to their units will be returned to them, no matter what happens on the battlefield or in what condition they return home. Tactical leaders understand the effects of protracted land operations on Soldiers and adjust the tempo of operations whenever circumstances allow. Senior commanders plan effective campaigns and major operations. They provide the resources needed to sustain operations, often through the imaginative use of joint capabilities.

1-76. The Army's preeminent challenge is to balance expeditionary agility and responsiveness with the endurance and adaptability needed to complete a campaign successfully, no matter what form it eventually takes. Landpower is a powerful complement to the global reach of American airpower and sea power. Prompt deployment of landpower gives joint force commanders options—for either deterrence or decisive action. Once deployed, landpower may be required for months or years. The initially deployed Army force will evolve constantly as the operational environment changes. Operational success depends on flexible employment of Army capabilities together with varying combinations of joint and interagency capabilities.

CLOSE COMBAT

1-77. Only on land do combatants come face-to-face with one another. Thus, the capability to prevail in close combat is indispensable and unique to land operations. It underlies most Army efforts in peace and war. *Close combat* **is warfare carried out on land in a direct-fire fight, supported by direct, indirect, and air-delivered fires**. Distances between combatants may vary from several thousand meters to hand-to-hand fighting. Close combat is required when other means fail to drive enemy forces from their positions. In that case, Army forces close with them and destroy or capture them. The outcome of battles and engagements depends on Army forces' ability to prevail in close combat. No other form of combat requires as much of Soldiers as it does.

1-78. Close combat is frequent in urban operations. **An** *urban operation* **is a military operation conducted where man-made construction and high population density are the dominant features**. The complexity of urban terrain and density of noncombatants reduce the effectiveness of advanced sensors and long-range and air-delivered weapons. Thus, a weaker enemy often attempts to negate Army advantages by engaging Army forces in urban environments. Operations in large, densely populated areas require special considerations. From a planning perspective, commanders view cities as both topographic features and dynamic entities containing hostile forces, local populations, and infrastructure. (JP 3-06 and FM 3-06 address these and other aspects of urban operations.)

UNCERTAINTY, CHANCE, AND FRICTION

> *Everything in war is very simple, but the simplest thing is difficult. The difficulties accumulate and end by producing a kind of friction that is inconceivable unless one has experienced war.... This tremendous friction, which cannot, as in mechanics, be reduced to a few points, is everywhere in contact with chance, and brings about effects that cannot be measured, just because they are largely due to chance.*
>
> Carl von Clausewitz
> *On War*[2]

1-79. Uncertainty, chance, and friction have always characterized warfare. On land, they are commonplace. Many factors inherent in land combat combine to complicate the situation. These include—

- Adverse weather.
- Chaos and confusion of battle.
- Complexity.
- Lack of accurate intelligence.
- Errors in understanding or planning.
- Fatigue.
- Misunderstanding among multinational partners.
- An adaptive and lethal enemy.
- Difficult terrain.
- Personality clashes.
- Civilian population.

1-80. Chance further complicates land operations. Things such as weather and other unforeseen events are beyond the control of a commander. For example, in December 1989, an ice storm at Fort Bragg, North Carolina, delayed deployment of some elements of the force invading Panama in Operation Just Cause. In addition to chance occurrences, enemy commanders have their own objectives and time schedules. These often lead to unforeseen encounters. Both enemy and friendly actions often produce unintended consequences, further complicating a situation, but they may lead to opportunities as well.

1-81. Several factors can reduce the effects of uncertainty, chance, and friction. Good leadership, flexible organizations, and dependable technology can lessen uncertainty. Timely, accurate intelligence may reduce the factors affected by chance. And a simple plan combined with continuous coordination might moderate the effects of friction. However, even when operations are going well, commanders make decisions based on incomplete, inaccurate, and contradictory information under adverse conditions. Determination is one means of overcoming friction; experience is another. High morale, sound organization, an effective command and control system, and well-practiced drills all help forces overcome adversity. Uncertainty, chance, and friction also affect the enemy, so commanders should look forward and exploit all opportunities. Understanding the operational environment, effective decisions, and flexibility in spite of adversity are essential to achieving tactical, operational, and strategic success.

COMPLEXITY

1-82. Future operational environments will be complex. While this does not necessarily equal a more dangerous environment, Soldiers can expect to deal with more complicated situations than ever before. The nature of land operations has expanded from a nearly exclusive focus on lethal combat with other armies to a complicated mixture of lethal and nonlethal actions directed at enemies, adversaries, and the local population, itself often a complicated mix. The enemy often follows no rules, while Army forces apply U.S. laws and international conventions to every conflict. The operational environment is saturated with information, with almost universal access to telecommunications and the Internet. The media will be ubiquitous. Action and message can no longer be separate aspects of operations because perception is so important to success. False reports, propaganda, rumors, lies, and inaccuracies spread globally faster than military authorities can correct or counter them, forcing Soldiers to deal with the consequences. Senior commanders and political leaders share tactical information in real time. Army forces work with and around a bewildering array of agencies and organizations—government, intergovernmental, nongovernmental, and commercial—and usually within a multinational military framework. American armed forces are the most advanced in the world. They have access to joint capabilities from the lowest echelons equaling unmatched combat power but at a cost in simplicity. Army forces will fight and operate in complex terrain and in cyberspace. These and many other factors increase the complexity of operations and stress every dimension of the Army's capabilities, especially the strength and depth of Army leaders.

SOLDIERS

1-83. Regardless of the importance of technological capabilities, it is Soldiers who accomplish the mission. Today's dangerous and complex security environment requires Soldiers who are men and women of character. Their character and competence represent the foundation of a values-based, trained, and ready Army. Soldiers train to perform tasks while operating alone or in groups. Soldiers and leaders develop the ability to exercise mature judgment and initiative under stress. The Army requires agile and adaptive leaders able to handle the challenges of full spectrum operations in an era of persistent conflict. Army leaders must be—

- Competent in their core proficiencies.
- Broad enough to operate across the spectrum of conflict.
- Able to operate in joint, interagency, intergovernmental, and multinational environments and leverage other capabilities in achieving their objectives.
- Culturally astute and able to use this awareness and understanding to conduct operations innovatively.
- Courageous enough to see and exploit opportunities in the challenges and complexities of the operational environment.
- Grounded in Army Values and the Warrior Ethos.

THE LAW OF WAR AND RULES OF ENGAGEMENT

1-84. Commanders at all levels ensure their Soldiers operate in accordance with the law of war. The *law of war* [also called the law of armed conflict] is that part of international law that regulates the conduct of armed hostilities (JP 1-02). It is the customary and treaty law applicable to the conduct of warfare on land and to relationships between belligerents and neutral states. The law of war includes treaties and international agreements to which the United States is a party as well as applicable customary international law. The purposes of the law of war are to—

- Protect both combatants and noncombatants from unnecessary suffering.
- Safeguard certain fundamental human rights of persons who become prisoners of war, the wounded and sick, and civilians.
- Make the transition to peace easier.

(FM 27-10 contains doctrine on the law of war.)

1-85. *Rules of engagement* are directives issued to guide United States forces on the use of force during various operations. These directives may take the form of execute orders, deployment orders, memoranda of agreement, or plans (JP 1-02). Rules of engagement always recognize a Soldier's inherent right of self-defense. Properly developed rules of engagement fit the situation and are clear, reviewed for legal sufficiency, and included in training. The joint staff and combatant commanders develop rules of engagement. The President and Secretary of Defense review and approve them. Rules of engagement vary between operations and may change during an operation. Adherence to them ensures Soldiers act consistently with international law, national policy, and military regulations.

1-86. The disciplined and informed application of lethal and nonlethal force is a critical contributor to successful Army operations and strategic success. All warfare, but especially irregular warfare, challenges the morals and ethics of Soldiers. An enemy may feel no compulsion to respect international conventions and indeed may commit atrocities with the aim of provoking retaliation in kind. Any loss of discipline on the part of Soldiers is then distorted and exploited in propaganda and magnified through the media. The ethical challenge rests heavily on small-unit leaders who maintain discipline and ensure that the conduct of Soldiers remains within ethical and moral boundaries. There are compelling reasons for this. First, humane treatment of detainees encourages enemy surrender and thereby reduces friendly losses. Conversely, nothing emboldens enemy resistance like the belief that U.S. forces will kill or torture prisoners. Second, humane treatment of noncombatants reduces their antagonism toward U.S. forces and may lead to valuable intelligence. Third, leaders make decisions in action fraught with consequences. If they lack an ethical foundation, those decisions become much, much harder. Finally, Soldiers must live with the consequences of their conduct. Every leader shoulders the responsibility that their subordinates return from a campaign not only as good Soldiers, but also as good citizens with pride in their service to the Nation.

1-87. The Soldier's Rules in AR 350-1 distill the essence of the law of war. They outline the ethical and lawful conduct required of Soldiers in operations. (Table 1-4 lists the Soldier's Rules.)

Table 1-4. The Soldier's Rules

- Soldiers fight only enemy combatants.
- Soldiers do not harm enemies who surrender. They disarm them and turn them over to their superior.
- Soldiers do not kill or torture enemy prisoners of war.
- Soldiers collect and care for the wounded, whether friend or foe.
- Soldiers do not attack medical personnel, facilities, or equipment.
- Soldiers destroy no more than the mission requires.
- Soldiers treat civilians humanely.
- Soldiers do not steal. Soldiers respect private property and possessions.
- Soldiers should do their best to prevent violations of the law of war.
- Soldiers report all violations of the law of war to their superior.

PREPARING AND TRAINING FOR FULL SPECTRUM OPERATIONS

1-88. Contemporary operations challenge the Army in many ways. The U.S. Army has always depended upon its ability to learn and adapt. German Field Marshal Erwin Rommel observed that American Soldiers were initially inexperienced but learned and adapted quickly and well. Even though the Army is much more experienced than it was in North Africa in World War II, today's complex environment requires organizations and Soldiers that can adapt equally quickly and well. To adapt, organizations constantly learn from experience (their own and that of others) and apply new knowledge to each situation. Flexibility and innovation are at a premium, as are creative and adaptive leaders. As knowledge increases, the Army will continuously adapt its doctrine, organization, training, materiel, leadership and education, personnel, and facilities.

1-89. The Army as a whole must be versatile enough to operate successfully across the spectrum of conflict from stable peace through insurgency to general war. Units must be agile enough to adapt quickly and be able to shift with little effort from a focus on one portion of the spectrum of conflict to focus on another. Change and adaptation that once required years to implement must now be recognized, communicated, and enacted far more quickly. Technology, having played an increasingly important role in increasing the lethality of the industrial age battlefield, will assume more importance and require greater and more rapid innovation in tomorrow's conflicts. No longer can responses to hostile asymmetric approaches be measured in months. Solutions must be fielded across the force in weeks—and then be adapted frequently and innovatively as the enemy adapts to counter the new-found advantages.

1-90. Leaders must be proficient in their core competencies, broad-minded enough to operate across the spectrum of conflict, and agile enough to adapt whenever necessary. They must be able to operate in joint, interagency, intergovernmental, and multinational environments leveraging political, economic, and informational efforts to achieve military objectives. And most importantly, leaders require an offensive mindset that focuses on the enemy and is opportunistic; they can see opportunities in challenges and act on them.

1-91. Effective training is the cornerstone of operational success. Through training, Soldiers, leaders, and units achieve the tactical and technical competence that builds confidence and allows them to conduct successful operations across the spectrum of conflict. The Army trains its forces using training doctrine that sustains their expeditionary and campaign excellence. Focused training prepares Soldiers, leaders, and units to deploy, fight, and win. This same training prepares Soldiers to create stable environments.

Achieving this competence requires specific, dedicated training on offensive, defensive, and stability or civil support tasks. The Army trains Soldiers and units daily in individual and collective tasks under challenging, realistic conditions. Training continues in deployed units to sustain skills and to adapt to changes in the operational environment.

1-92. The United States' responsibilities are global; therefore, Army forces prepare to operate in any environment. Army training develops confident, competent, and agile leaders and units. Training management, including mission-essential task list development, links training with missions. Commanders focus their training time and other resources on tasks linked to their mission. Because Army forces face diverse threats and mission requirements, senior commanders adjust their training priorities based on the likely operational environment. As units prepare for deployment, commanders adapt training priorities to address tasks required by actual or anticipated operations. (FMs 7-0 and 7-1 describe training management.)

1-93. Army training includes a system of techniques and standards that allow Soldiers and units to determine, acquire, and practice necessary skills. Candid assessments, after action reviews, and applying lessons learned and best practices produce quality Soldiers and versatile units, ready for all aspects of the situation. The Army's training system prepares Soldiers and leaders to employ Army capabilities adaptively and effectively in today's varied and challenging conditions.

1-94. Through training, the Army prepares Soldiers to win in land combat. Training builds teamwork and cohesion within units. It recognizes that Soldiers ultimately fight for one another and their units. Training instills discipline. It conditions Soldiers to operate within the law of war and rules of engagement. Training prepares unit leaders for the harsh reality of land combat. It emphasizes the fluid and disorderly conditions inherent in land operations.

1-95. Within these training situations, commanders emphasize mission command. To employ mission command successfully during operations, units must understand, foster, and frequently practice its principles during training. (Paragraphs 3-29 through 3-35 discuss mission command.)

SUMMARY

1-96. Successful mission accomplishment requires understanding the operational environment, the role of the Army in unified action, and how Soldiers, leaders, and units accomplish missions through full spectrum operations. Army forces conduct prompt and sustained operations as part of a joint force to conclude hostilities and establish conditions favorable to the host nation, the United States, and their multinational partners. Today's operational environments are complex and require continuous learning and adaptation. Commanders use experience, applied judgment, and various analytic tools to gain the situational understanding necessary to make timely decisions to maintain the initiative and achieve decisive results. The more commanders understand their operational environment, the more effectively they can employ forces.

This page intentionally left blank.

Chapter 2

The Continuum of Operations

The continuum of operations frames the application of landpower. It includes the spectrum of conflict and the operational themes. The spectrum of conflict is an ascending scale of violence ranging from stable peace to general war. Operational themes give commanders a way to characterize the dominant major operation underway in an area of operations. The themes also provide overlapping categories for grouping types of operations from the land force perspective. The continuum of operations thus links the operational environment, discussed in chapter 1, with the Army's operational concept—full spectrum operations—discussed in chapter 3.

THE SPECTRUM OF CONFLICT

2-1.   The spectrum of conflict is the backdrop for Army operations. It places levels of violence on an ascending scale marked by graduated steps. (See figure 2-1.) The spectrum of conflict spans from stable peace to general war. It includes intermediate levels of unstable peace and insurgency. In practice, violent conflict does not proceed smoothly from unstable peace through insurgency to general war and back again. Rather, general war and insurgencies often spark additional violence within a region, creating broad areas of instability that threaten U.S. vital interests. Additionally, the level of violence may jump from one point on the spectrum to another. For example, unstable peace may erupt into general war, or general war may end abruptly in unstable peace. Therefore, the four levels are not an exclusive set. Nonetheless, the spectrum of conflict provides a tool to understand and visualize the level of violence and the corresponding role of the military in resolving a conflict.

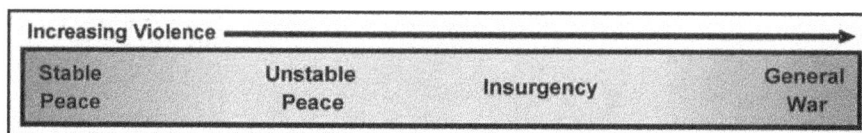

Figure 2-1. The spectrum of conflict

2-2.   Army forces affect the operational environment as they operate to accomplish their mission. Commanders seek to establish conditions favorable for conducting subsequent operations and tasks. They consider not only their follow-on missions but also how to restore a stable peace. If stable peace is not readily attainable, commanders design operations so that in the end, they lower the violence level as much as possible. When operating anywhere on the spectrum, commanders and staffs consider how to move the level of violence toward stable peace.

2-3.   Military power cannot, by itself, restore or guarantee stable peace. It must, however, establish global, regional, and local conditions that allow the other instruments of national power—diplomatic, informational, and economic—to exert their full influence. For example, the Nation's air, land, and sea power deter threats posed by hostile powers by holding their homeland and vital interests at risk. This creates avenues for diplomacy to resolve disputes. Multinational forces may separate warring factions to stop a civil war that threatens regional peace Their actions allow international aid organizations to reach masses of refugees and an international commission to seek some sort of equitable settlement. On a local level, an Army task force suppresses terrorism and lawlessness so other government agencies can work freely with host-nation officials to restore self-sustaining governance. In each case, achieving stable peace requires expertise and capabilities beyond those developed in the military force. Every use of military force to restore stable peace requires the other three instruments of national power. Conversely, diplomatic, informational, and economic efforts to restore a stable peace are usually futile unless backed by effective military power, military power with global reach and endurance. In every campaign and major operation, success, as characterized by a stable peace, depends on unified action involving concerted efforts by multinational military and civilian partners.

STABLE PEACE

2-4.   Stable peace is characterized by the absence of militarily significant violence. Activities of international actors (such as states, corporations, and nongovernmental organizations) are confined to peaceful interaction in politics, economics, and other areas of interest. Peaceful interaction may include intense competition as well as cooperation and assistance. While tensions do exist, all recognize that their interests are best achieved by means other than violence.

UNSTABLE PEACE

2-5.   When one or more parties threaten or use violence to achieve their objectives, stable peace degenerates into unstable peace. Unstable peace may also result when violence levels decrease after violent conflict. In some cases, outside powers may apply force to limit conflict. Preventing a return to violent conflict may require peace operations. (See paragraphs 2-34 through 2-44.) Sometimes stable peace is not immediately achievable. At those times, the goal of conflict termination is establishing conditions in which peace operations can prevent conflict from recurring. Doing this allows the instruments of national power to work toward stable peace.

INSURGENCY

2-6.   Joint doctrine defines an *insurgency* as an organized movement aimed at the overthrow of a constituted government through use of subversion and armed conflict (JP 1-02). It is a condition of politically motivated conflict involving significant intra- or interstate violence but usually short of large-scale operations by opposing conventional forces. Insurgencies often include widespread use of irregular forces and terrorist tactics. An insurgency may develop in the aftermath of general war or through degeneration of unstable peace. Insurgencies may also emerge on their own from chronic social or economic conditions. In addition, some conflicts, such as the Chinese Revolution, have escalated from protracted insurgencies into general wars. Intervention by a foreign power in an insurgency may increase the threat to regional stability.

GENERAL WAR

2-7.   *General war* is armed conflict between major powers in which the total resources of the belligerents are employed, and the national survival of a major belligerent is in jeopardy (JP 1-02). General war usually involves nation-states and coalitions; however, civil wars may reach this level of violence. In general war, large and heavily armed conventional forces fight for military supremacy by conducting major combat operations. These operations aim to defeat the enemy's armed forces and eliminate the enemy's military capability. These conflicts are dominated by large-scale conventional operations but often include guerrilla and unconventional warfare. To illustrate, Soviet partisans waged unconventional warfare against German lines of communications during World War II. The Vietcong conducted guerrilla warfare throughout the Vietnam War, even as the North Vietnamese Army fought conventional battles against U.S. and South Vietnamese forces.

ARMY FORCES AND THE SPECTRUM OF CONFLICT

2-8.   Army forces operate anywhere on the spectrum of conflict. In each case, achieving the end state requires reducing the violence level and creating conditions that advance U.S. national strategic goals. Commanders conduct a series of operations intended to establish conditions conducive to a stable peace. Some situations require applying massive force in major combat operations to eliminate a threat; others involve applying military power to reduce an insurgency to a size the host-nation forces can defeat. The goal at any point is to move conditions to a lower level of violence; however, avoiding intermediate levels is desirable. When this is not possible, commanders seek to move the situation through them to stable peace as quickly as possible.

2-9.   Today's operational environment requires Army forces to continuously evaluate and adapt their tactics to ensure that they are appropriate. Recent experience demonstrates the difficulty and cost of fighting terrorists and insurgents while supporting reconstruction efforts. These experiences and a study of other conflicts have revealed insights that are guiding the Army's effort to prepare for future operations:

- All major operations combine offensive, defensive, and stability elements executed simultaneously at multiple echelons.
- The operational environment evolves over time and changes due to military operations.
- Operations conducted during one phase of a campaign or major operation directly affect subsequent phases. Commanders should conduct current operations in a manner that sets the conditions necessary for future operations—and ultimately allowing the other instruments of national power to secure a stable peace.
- Major operations are conducted not only to defeat the enemy but also to restore a stable peace. The military plays a large role in this effort, even after major combat operations have ended. Restoring a stable peace after a violent conflict may take longer and be more difficult than defeating enemy forces.
- In any campaign or major operation, changing conditions require Army forces to adapt their tactics, techniques, and procedures to the operational environment. To be successful, leaders must develop learning organizations that collect and share best practices and lessons learned.

OPERATIONAL THEMES

2-10.   Army forces conduct major operations to defeat an enemy and to establish conditions necessary to achieve the national strategic end state. A *major operation* is a series of tactical actions (battles, engagements, strikes) conducted by combat forces of a single or several Services, coordinated in time and place, to achieve strategic or operational objectives in an operational area. These actions are conducted simultaneously or sequentially in accordance with a common plan and are controlled by a single commander. For noncombat operations, [a major operations refers] to the relative size and scope of a military operation (JP 3-0). Examples of major operations include Operation Cobra (the breakout from the Normandy beachhead during World War II) and Operation Chromite (the amphibious landing at Inchon during the Korean War).

2-11.   Conflict intensity varies over time and among locations; therefore, it is difficult to precisely describe a major operation's character. In fact, the character of most major operations is likely to evolve. All major operations comprise many smaller operations conducted simultaneously. These also may vary with time. Nevertheless, it is possible to establish a theme for each major operation, one that distinguishes it from other operations with different characteristics. Major combat operations, for instance, differ distinctly from counterinsurgency operations; both differ from peace operations. Different themes usually demand different approaches and force packages, although some activities are common to all.

2-12.   **An *operational theme* describes the character of the dominant major operation being conducted at any time within a land force commander's area of operations. The operational theme helps convey the nature of the major operation to the force to facilitate common understanding of how the commander broadly intends to operate.** Operational themes have implications for task-organization, resource allocation, protection, and tactical task assignment. They establish a taxonomy for understanding the kinds of joint and Army major operations and relationships among them.

2-13.   Grouping military operations with common characteristics under operational themes allows doctrine to be developed for each theme rather than for a multitude of joint operations. (See table 2-1, page 2-4.) However, this taxonomy does not limit when commanders may use a type of operation. Some operations listed under one operational theme are routinely conducted within major operations characterized by another. For example, noncombatant evacuation operations may be conducted during counterinsurgency, or support to an insurgency may occur during major combat operations. Such situations do not change the broader character of the major operation. The operational themes emphasize the differences among the various types of joint operations. These differences are usually greater for land forces (including special operations forces) than for the other Services.

Table 2-1. Examples of joint military operations conducted within operational themes

Peacetime military engagement	Limited intervention	Peace operations	Irregular warfare
• Multinational training events and exercises • Security assistance • Joint combined exchange training • Recovery operations • Arms control • Counterdrug activities	• Noncombatant evacuation operations • Strike • Raid • Show of force • Foreign humanitarian assistance • Consequence management • Sanction enforcement • Elimination of weapons of mass destruction	• Peacekeeping • Peace building • Peacemaking • Peace enforcement • Conflict prevention	• Foreign internal defense • Support to insurgency • Counterinsurgency • Combating terrorism • Unconventional warfare
Note: Major combat operations usually involve a series of named major operations, such as Operation Desert Storm, each involving significant offensive and defensive operations and supporting air, land, sea, and special operations.			

2-14. Each operational theme corresponds broadly to a range along the spectrum of conflict. (See figure 2-2.) Commanders describe their desired end state as how they envision the conditions of the operational environment when the operation ends. Often, this envisions either a condition of stable peace or at least one in which civilian organizations can build toward a stable peace with a sustainable military commitment. A stable peace may include any or all of the following: a safe and secure populace, a legitimate central government, a viable market economy, and an effective rule of law. Lacking involvement by civilian organizations, none of these goals is sustainable. To ensure progress toward the end state, higher level commanders continuously assess the overall campaign and their subordinates' operations. They adjust the type of operation as each campaign phase unfolds. This adjustment affects their focus, resource allocation, and directed tasks. Commanders visualize how they see the conflict ending and prepare to transition between operational themes as the operation progresses. They pursue avenues for increased cooperation with civilian agencies, handing over activities to civilian direction as soon as conditions permit.

2-15. Operational themes provide a useful means of characterizing phases of a campaign. The transition between operational themes requires careful planning and continuous assessment. For example, at the conclusion of major combat operations, the character of the campaign may evolve to irregular warfare or peace operations. While the scope of their defeat may induce an enemy to accept occupation and peace enforcement without a period of irregular warfare, commanders plan for a potential insurgency and prepare accordingly. Shifting from one operational theme to another often requires adjustments to the composition of the force. These adjustments apply not only to task-organizing the force but to deploying and redeploying units, establishing new bases, and dismantling bases no longer needed. In particular, a change in operational theme may require modification to the mission-essential task lists and additional training for both deploying units and units in theater. The responsibilities between military commanders and other government officials may also change. For example, the ambassador may become the senior U.S. Government official as opposed to the joint force commander.

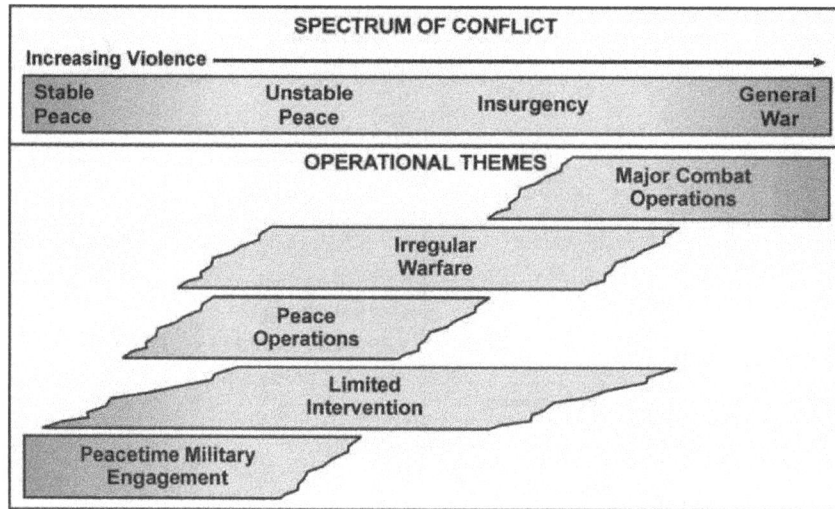

Figure 2-2. The spectrum of conflict and operational themes

2-16. Operational themes should not be confused with tactical tasks or activities. Operational themes are too general to be assigned as missions. Rather, they describe the major operation's general characteristics, not the details of its execution. The theme of a major operation may change for various reasons such as—

- Planned phases.
- Changes caused by friendly, enemy, or neutral activity.
- Revised political guidance.
- Unexpected opportunities.

PEACETIME MILITARY ENGAGEMENT

2-17. *Peacetime military engagement* **comprises all military activities that involve other nations and are intended to shape the security environment in peacetime. It includes programs and exercises that the United States military conducts with other nations to shape the international environment, improve mutual understanding, and improve interoperability with treaty partners or potential coalition partners. Peacetime military engagement activities are designed to support a combatant commander's objectives within the theater security cooperation plan.** Peacetime military engagement encourages regional stability. These activities may be long term, such as training teams and advisors assisting land forces, or short term, such as multinational exercises. Combat is not envisioned, although terrorist attacks against deployed forces are always possible. Policy, regulations, and security cooperation plans, rather than doctrine, typically govern peacetime military engagement activities. These are usually conducted bilaterally but can involve multiple nations. Examples of joint operations and activities that fall under peacetime military engagement include the following:

- Multinational training events and exercises.
- Security assistance.
- Joint combined exchange training.
- Recovery operations.
- Arms control.
- Counterdrug activities.

Multinational Training Events and Exercises

2-18. Combatant commanders support many multinational training events and exercises. These exercises have many purposes, including—

- Demonstrating military capabilities to potential aggressors.
- Improving interoperability.
- Establishing or improving military-to-military ties with another nation.

Security Assistance

2-19. The security assistance program is an important instrument of U.S. foreign and national security policy. It contributes to—

- Deterrence.
- Promoting regional stability.
- Guaranteeing U.S. access to vital overseas military facilities.
- Increasing host-nation military capabilities, thus reducing unilateral U.S. military requirements.
- Enhancing weapons standardization and interoperability in support of multinational force compatibility. (AR 34-1 governs multinational force compatibility.)
- Supporting the U.S. defense industrial base.

2-20. Security assistance includes various supporting programs for foreign military logistic, financial, and military education assistance. For example, the international military education and training program is part of security assistance. It also includes U.S. military teams sent to multinational partners to assist in training. (AR 12-1 governs security assistance.)

Joint Combined Exchange Training

2-21. *Joint combined exchange training* is a program conducted overseas to fulfill U.S. forces training requirements and at the same time exchange the sharing of skills between U.S. forces and host-nation counterparts. Training activities are designed to improve U.S. and host-nation capabilities (JP 3-05). Army forces routinely participate in exchange programs, often at the small-unit level. Special operations forces use this program to improve their regional expertise while contributing to a combatant commander's theater security cooperation plan. (JP 3-05 addresses joint combined exchange training.)

Recovery Operations

2-22. *Recovery operations* are operations conducted to search for, locate, identify, recover, and return isolated personnel, human remains, sensitive equipment, or items critical to national security (JP 3-50). (JP 3-50 and FM 3-50.1 contain doctrine on personnel recovery.)

Arms Control

2-23. Combatant commanders support multinational arms control agreements concerning prohibited weapons and illegal arms trafficking. They also provide forces and control means to block the sale or transfer of arms to terrorists or other criminals as the Secretary of Defense directs. Such actions may be unilateral or multinational.

Counterdrug Activities

2-24. *Counterdrug activities* are those measures taken to detect, interdict, disrupt, or curtail any activity that is reasonably related to illicit drug trafficking. This includes, but is not limited to, measures taken to detect, interdict, disrupt, or curtail activities related to substances, materiel, weapons, or resources used to finance, support, secure, cultivate, process, or transport illegal drugs (JP 3-07.4). (JP 3-07.4 contains doctrine on counterdrug activities.)

LIMITED INTERVENTION

2-25. Limited interventions are executed to achieve an end state that is clearly defined and limited in scope. Corresponding limitations are imposed on the supporting operations and size of the forces involved. These operations may be phased but are not intended to become campaigns. Although limited interventions are confined in terms of end state and forces, their execution may be lengthy. Joint task forces usually conduct limited interventions. The most common types of limited interventions are the following:

- Noncombatant evacuation operations.
- Strike.
- Raid.
- Show of force.
- Foreign humanitarian assistance.
- Consequence management.
- Sanction enforcement.
- Elimination of weapons of mass destruction.

Noncombatant Evacuation Operations

2-26. *Noncombatant evacuation operations* are operations directed by the Department of State or other appropriate authority, in conjunction with the Department of Defense, whereby noncombatants are evacuated from foreign countries when their lives are endangered by war, civil unrest, or natural disaster to safe havens or to the United States (JP 3-0). (JP 3-68 contains doctrine for noncombatant evacuation operations.)

Strike

2-27. A *strike* is an attack to damage or destroy an objective or a capability (JP 3-0). While strikes are conducted as part of tactical operations, in limited interventions they are conducted by a joint force apart from a campaign or major operation. An example of a strike conducted as a limited intervention is Operation El Dorado Canyon, executed in 1986. It consisted of a series of air strikes on targets inside Libya.

Raid

2-28. A *raid* is an operation to temporarily seize an area in order to secure information, confuse an adversary, capture personnel or equipment, or to destroy a capability. It ends with a planned withdrawal upon completion of the assigned mission (JP 3-0). Raids are routinely conducted as part of tactical operations but may be conducted as separate joint operations. The latter is characterized as a limited intervention. (FM 3-90 contains doctrine on tactical-level raids.)

Show of Force

2-29. A *show of force* is an operation designed to demonstrate U.S. resolve that involves increased visibility of U.S. deployed forces in an attempt to defuse a specific situation that, if allowed to continue, may be detrimental to U.S. interests or national objectives (JP 3-0).

Foreign Humanitarian Assistance

2-30. *Foreign humanitarian assistance* consists of programs conducted to relieve or reduce the results of natural or man-made disasters or other endemic conditions such as human pain, disease, hunger, or privation that might present a serious threat to life or that can result in great damage to or loss of property. Foreign humanitarian assistance provided by U.S. forces is limited in scope and duration. The foreign assistance provided is designed to supplement or complement the efforts of the host-nation civil authorities or agencies that may have the primary responsibility for providing foreign humanitarian assistance. Foreign humanitarian assistance operations are those conducted outside the United States, its territories, and possessions (JP 3-33). An example of foreign humanitarian assistance is the multinational relief operation sent to Indonesia after the December 2004 tsunami. (JP 3-07.6 contains doctrine for foreign humanitarian assistance.)

Consequence Management

2-31. *Consequence management* involves actions taken to maintain or restore essential services and manage and mitigate problems resulting from disasters and catastrophes, including natural, man-made, or terrorist incidents (JP 1-02). (JP 3-41 contains doctrine on chemical, biological, radiological, nuclear, and high-yield explosives consequence management.)

Sanction Enforcement

2-32. *Sanction enforcement* comprises operations that employ coercive measures to interdict the movement of certain types of designated items into or out of a nation or specified area (JP 3-0).

Elimination of Weapons of Mass Destruction

2-33. Operations to eliminate weapons of mass destruction systematically locate, characterize, secure, disable, or destroy a state or nonstate actor's weapons of mass destruction and related capabilities. Elimination operations are one of eight joint mission areas—offensive operations, elimination operations, interdiction operations, active defense, passive defense, weapons of mass destruction consequence management, security cooperation, and threat reduction. These areas make up the three pillars (nonproliferation, counterproliferation, and consequence management) of combating weapons of mass destruction. (JP 3-40 contains doctrine on combating weapons of mass destruction.)

PEACE OPERATIONS

2-34. *Peace operations* is a broad term that encompasses multiagency and multinational crisis response and limited contingency operations involving all instruments of national power with military missions to contain conflict, redress the peace, and shape the environment to support reconciliation and rebuilding and facilitate the transition to legitimate governance. Peace operations include peacekeeping, peace enforcement, peacemaking, peace building, and conflict prevention efforts (JP 3-07.3). Army forces conduct the following types of peace operations:

- Peacekeeping.
- Peace building.
- Peacemaking.
- Peace enforcement.
- Conflict prevention.

2-35. The objectives of peace operations include keeping violence from spreading, containing violence that has occurred, and reducing tension among factions. Accomplishing these objectives creates an environment in which other instruments of national power are used to reduce the level of violence to stable peace. Peace operations are usually interagency efforts. They require a balance of military and diplomatic resources. (JP 3-07.3 and FM 3-07 contain doctrine for peace operations.)

Peacekeeping

2-36. *Peacekeeping* consists of military operations undertaken with the consent of all major parties to a dispute, designed to monitor and facilitate implementation of an agreement (cease fire, truce, or other such agreement) and support diplomatic efforts to reach a long-term political settlement (JP 3-07.3).

Peace Building

2-37. *Peace building* involves stability actions, predominately diplomatic and economic, that strengthen and rebuild governmental infrastructure and institutions in order to avoid a relapse into conflict (JP 3-0). Peace building provides the reconstruction and societal rehabilitation in the aftermath of conflict that offers hope to the host-nation populace. Stability operations promote reconciliation, strengthen and rebuild civil infrastructures and institutions, build confidence, and support economic reconstruction to prevent a return to conflict. The ultimate measure of success in peace building is political, not military.

Peacemaking

2-38. *Peacemaking* is the process of diplomacy, mediation, negotiation, or other forms of peaceful settlements that arranges an end to a dispute and resolves issues that led to it (JP 3-0).

Peace Enforcement

2-39. *Peace enforcement* involves the application of military force, or the threat of its use, normally pursuant to international authorization, to compel compliance with resolutions or sanctions designed to maintain or restore peace and order (JP 3-0).

Conflict Prevention

2-40. Conflict prevention consists of actions taken before a predictable crisis to prevent or limit violence, deter parties, and reach an agreement before armed hostilities begin. Conflict prevention often involves diplomatic initiatives. It also includes efforts designed to reform a country's security sector and make it more accountable to civilian control. Conflict prevention may require deploying forces to contain a dispute or prevent it from escalating into hostilities. (JP 3-07.3 contains doctrine on conflict prevention.)

Considerations for Peace Operations

2-41. Peace operations ease the transition to a stable peace in a war-torn nation or region by supporting reconciliation and rebuilding. They are often conducted under international supervision. U.S. forces may conduct peace operations under the sponsorship of the United Nations, another intergovernmental organization, as part of a coalition, or unilaterally.

2-42. Peace operations are often conducted in complex, ambiguous, and uncertain environments. The operational environment for a peace operation may include any or all of the following characteristics:

- Asymmetric threats.
- Failing or failed states.
- Absence of the rule of law.
- Terrorism and terrorist organizations.
- Gross violations of human rights.
- Collapse of civil infrastructure.
- Presence of dislocated civilians.

2-43. Army forces in peace operations strive to create a safe and secure environment, primarily through stability operations. Army forces use their offensive and defensive capabilities to deter external and internal adversaries from overt actions against each other. Establishing security and control enables civilian agencies to address the underlying causes of the conflict and generate a self-sustaining peace. Army forces provide specialized support to other government agencies as necessary.

2-44. Peace operations require opposing parties to cooperate with the international community. In most peace operations, this comes voluntarily. However, peace enforcement involves the threat or use of military force to compel cooperation. Successful peace operations also require support from the local populace and host-nation leaders. There is less likelihood of combat, and when it occurs, it is usually at the small-unit level. Therefore, small-unit operations are important to success. Units involved in peace operations prepare for sudden engagements, even while executing operations to prevent them. Commanders emphasize the use of information, particularly information used to inform and influence various opposing audiences in the area of operations. (See paragraphs 7-10 through 7-22.) Peace operations require perseverance to achieve the desired end state.

IRREGULAR WARFARE

2-45. *Irregular warfare* **is a violent struggle among state and nonstate actors for legitimacy and influence over a population**. This broad form of conflict has insurgency, counterinsurgency, and unconventional warfare as the principal activities. Irregular forces are normally active in these conflicts. However, conventional forces may also be heavily involved, particularly in counterinsurgencies.

2-46. Irregular warfare differs from conventional operations dramatically in two aspects. First, it is warfare among and within the people. The conflict is waged not for military supremacy but for political power. Military power can contribute to the resolution of this form of warfare, but it is not decisive. The effective application of military forces can create the conditions for the other instruments of national power to exert their influence. Secondly, irregular warfare also differs from conventional warfare by its emphasis on the indirect approach. (See paragraph 6-41.) Irregular warfare avoids a direct military confrontation. Instead, it combines irregular forces and indirect, unconventional methods (such as terrorism) to subvert and exhaust the opponent. It is often the only practical means for a weaker opponent to engage a powerful military force. Irregular warfare seeks to defeat the opponent's will through steady attrition and constant low-level pressure. In some instances, it targets the populace and avoids conventional forces. This approach creates instability. It severely challenges civil authority to fulfill its first responsibility—providing security.

2-47. Special operations forces conduct most irregular warfare operations. Sometimes conventional forces support them; other times special operations forces operate alone. However, if special operations forces and host-nation forces cannot defeat unconventional and irregular threats, conventional Army forces may assume the lead role. The joint operations grouped under irregular warfare include the following:

- Foreign internal defense.
- Support to insurgency.
- Counterinsurgency.
- Combating terrorism.
- Unconventional warfare.

Foreign Internal Defense

2-48. *Foreign internal defense* is the participation by civilian and military agencies of a government in any of the action programs taken by another government or other designated organization to free and protect its society from subversion, lawlessness, and insurgency (JP 3-05). The categories of foreign internal defense operations are indirect support and direct support.

Indirect Support

2-49. Indirect support emphasizes host-nation self-sufficiency. It builds strong national infrastructures through economic and military capabilities. Examples include security assistance programs, multinational exercises, and exchange programs. Indirect support reinforces host-nation legitimacy and primacy in addressing internal problems by keeping U.S. military assistance inconspicuous.

Direct Support

2-50. Direct support uses U.S. forces to assist the host-nation civilian populace or military forces directly. Direct support includes operational planning assistance, civil affairs activities, intelligence and communications sharing, logistics, and training of local military forces. It may also involve limited combat operations, usually in self-defense.

Considerations for Foreign Internal Defense

2-51. Foreign internal defense involves all instruments of national power. It is primarily a series of programs that supports friendly nations operating against or threatened by hostile elements. Foreign internal defense promotes regional stability by helping a host nation respond to its people's needs while maintaining security. Participating Army forces normally advise and assist host-nation forces while refraining from combat operations.

2-52. Foreign internal defense is a significant mission for selected Army special operations forces. However, it requires joint planning, preparation, and execution to integrate and focus the efforts of all Service and functional components. These missions are approved by the President, limited in scope and duration, and conducted in support of legitimate host-nation forces.

2-53. Foreign internal defense operations often respond to growing insurgencies. Most of these activities help a host nation prevent an active insurgency from developing further. If an insurgency already exists or preventive measures fail, foreign internal defense focuses on using host-nation security forces and other resources to eliminate, marginalize, or assimilate insurgent elements. The United States provides military support to host-nation counterinsurgency efforts, recognizing that military power alone cannot achieve lasting success. Host-nation and U.S. actions promote a secure environment with programs that eliminate causes of insurgencies. Military support to a threatened government balances security with economic development to enhance or reestablish stability. (JP 3-07.1 and FM 3-05.202 contain doctrine for foreign internal defense.)

Support to Insurgency

2-54. Army forces may support insurgencies against regimes that threaten U.S. interests. Normally Army special operations forces provide the primary U.S. land forces. These forces' training, organization, and regional focus make them well suited for these operations. Conventional Army forces that support insurgencies provide logistic and training support but normally do not conduct offensive or defensive operations.

Counterinsurgency

2-55. *Counterinsurgency* is those military, paramilitary, political, economic, psychological, and civic actions taken by a government to defeat insurgency (JP 1-02). In counterinsurgency, host-nation forces and their partners operate to defeat armed resistance, reduce passive opposition, and establish or reestablish the host-nation government's legitimacy. Counterinsurgency is the dominant joint operation in Operations Iraqi Freedom and Enduring Freedom. (FM 3-24 discusses counterinsurgency.)

2-56. Insurgents try to persuade the populace to accept the insurgents' goals or force political change. When persuasion does not work, insurgents use other methods to achieve their goals. These may include intimidation, sabotage and subversion, propaganda, terror, and military pressure. Sometimes insurgents attempt to organize the populace into a mass movement. At a minimum, they aim to make effective host-nation governance impossible. Some insurgencies are transnational. Other situations involve multiple insurgencies underway in an area at the same time. Counterinsurgency becomes more complex in these situations.

2-57. While each insurgency is unique, similarities among them exist. Insurgencies are more likely to occur in states with a lack of national cohesion or with weak, inefficient, unstable, or unpopular governments. Internal conflicts may be racial, cultural, religious, or ideological. Additional factors, such as corruption and external agitation, may also fuel an insurgency. Successful insurgencies develop a unifying leadership and organization and an attractive vision of the future. Usually only insurgencies able to attract widespread, popular support pose a real threat to state authority.

2-58. Most operations in counterinsurgencies occur at the small-unit level—squad, platoon, or company. However, larger operations also occur, and a consistent, long-range plan is essential to defeat an insurgency. Commanders carefully assess the negative effects of violence on the populace. Strict adherence to the rules of engagement is essential. Operations should reflect and promote the host-nation government's authority. This undermines insurgent attempts to establish an alternative authority. It also reduces the tendency of the population to view the units conducting counterinsurgency as an occupying force.

2-59. Larger units, such as brigades and divisions, provide direction and consistency to Army operations in their areas of operations and mass resources and forces to make operations more effective. They also respond to any threat large enough to imperil the smaller units distributed throughout the area of operations.

Lower echelons can then operate across larger areas (against rural insurgencies) or among greater populations (against urban insurgencies).

Combating Terrorism

2-60. *Combating terrorism* is actions, including antiterrorism (defensive measures taken to reduce vulnerability to terrorist acts) and counterterrorism (offensive measures taken to prevent, deter, and respond to terrorism), taken to oppose terrorism throughout the entire threat spectrum (JP 3-07.2). *Terrorism* is the calculated use of unlawful violence or threat of unlawful violence to inculcate fear; [these acts are] intended to coerce or to intimidate governments or societies in the pursuit of goals that are generally political, religious, or ideological (JP 3-07.2). An enemy who cannot defeat conventional Army forces may resort to terrorism. Terrorist attacks can create disproportionate effects on conventional forces. Their effect on societies can be even greater. Terrorist tactics may range from individual assassinations to employing weapons of mass destruction.

Counterterrorism

2-61. *Counterterrorism* is operations that include the offensive measures taken to prevent, deter, preempt, and respond to terrorism (JP 3-05). Counterterrorism actions include strikes and raids against terrorist organizations and facilities outside the United States and its territories. Although counterterrorism is a specified mission for selected special operations forces, conventional Army forces may also contribute. Commanders who employ conventional forces against terrorists are conducting offensive operations, not counterterrorism operations.

Antiterrorism

2-62. *Antiterrorism* is defensive measures used to reduce the vulnerability of individuals and property to terrorist acts, to include limited response and containment by local military and civilian forces (JP 3-07.2). It is a protection task. All forces consider antiterrorism during all operations. Commanders take the security measures necessary to accomplish the mission and protect their forces against terrorism. They make every reasonable effort to minimize their forces' vulnerability to violence and hostage taking. Typical antiterrorism actions include—

- Completing unit and installation threat and vulnerability assessments.
- Training in antiterrorism awareness.
- Establishing special reaction teams and protective services at installations and bases.
- Ensuring that antiterrorism measures protect personnel, physical assets, and information, including high-risk personnel and designated critical assets.
- Establishing civil-military partnerships for weapons of mass destruction crises and consequence management.
- Developing terrorist threat and incident response plans that include managing the force protection condition system.
- Establishing appropriate policies based on the threat and force protection condition system.

(JP 3-07.2 contains doctrine for antiterrorism.)

Unconventional Warfare

2-63. *Unconventional warfare* is a broad spectrum of military and paramilitary operations, normally of long duration, predominantly conducted through, with, or by indigenous or surrogate forces who are organized, trained, equipped, supported, and directed in varying degrees by an external source. It includes, but is not limited to, guerrilla warfare, subversion, sabotage, intelligence activities, and unconventional assisted recovery (JP 3-05). Within the U.S. military, conduct of unconventional warfare is a highly specialized special operations force mission. Special operations forces may conduct unconventional warfare as part of a separate operation or within a campaign. During Operation Enduring Freedom, special operations forces and other government agencies conducted unconventional warfare within the joint campaign to topple the Taliban regime.

2-64. Conventional Army forces may support unconventional warfare. For example, during Operation Iraqi Freedom, conventional forces supported Joint Task Force-North by securing bases in the joint special operations area. (JP 3-05 contains doctrine on unconventional warfare conducted by Army special operations forces.)

MAJOR COMBAT OPERATIONS

2-65. Major combat operations occur in circumstances usually characterized as general war. States, alliances, or coalitions usually resort to war because significant national or multinational interests are threatened. Combat between large formations characterizes these operations. Major combat operations conducted by U.S. forces are always joint operations, although an Army headquarters may form the base of a joint force headquarters. These operations typically entail high tempo, high resource consumption, and high casualty rates.

2-66. Major combat operations often include combat between the uniformed armed forces of nation-states. Even then, these operations tend to blur with other operational themes. For example, in Vietnam both the United States and North Vietnam deployed their national armed forces and, although major battles occurred, the United States characterized much of the war as counterinsurgency.

2-67. Civil wars, particularly within a developed nation, often include major combat operations. The American Civil War, the Russian Revolution, and Yugoslavia's collapse all involved recurring, high-intensity clashes between armies. Even in less developed regions, civil war leads to massive casualties among combatants and noncombatants alike. Insurgencies can develop into civil wars, particularly when external powers back both the government and the insurgents.

2-68. Not all major combat operations are protracted. Joint operations may capitalize on superior military capability to quickly overwhelm a weaker enemy. Examples include the following:
- The coup de main in Panama in 1989.
- The forcible entry of Grenada in 1983.
- The major combat operations against Iraq in 1991 and 2003.

2-69. Successful major combat operations defeat or destroy the enemy's armed forces and seize terrain. Commanders assess them in terms of numbers of military units destroyed or rendered combat ineffective, the level of enemy resolve, and the terrain objectives seized or secured. Major combat operations are the operational theme for which doctrine, including the principles of war, was originally developed.

SUMMARY

2-70. Commanders use the spectrum of conflict to describe the level of violence in terms of an ascending scale marked by graduated steps. Army forces operate anywhere along the spectrum of conflict, from peacetime military engagement in areas of stable peace to major combat operations during general war. In each case, the objective is to create conditions that advance U.S. goals. The operational themes group types of military operations according to common characteristics. Operational themes establish a taxonomy for understanding the many kinds of joint operations and the relationships among them. Commanders convey the overall character of a major operation, including the principles that govern it, in terms of its operational theme. The operational theme varies during a campaign or major operation. Although many tactical activities are common to all, different themes demand different approaches.

This page intentionally left blank.

Chapter 3

Full Spectrum Operations

The foundations for Army operations are contained in its operational concept—full spectrum operations. The goal of full spectrum operations is to apply landpower as part of unified action to defeat the enemy on land and establish the conditions that achieve the joint force commander's end state. The complexity of today's operational environments requires commanders to combine offensive, defensive, and stability or civil support tasks to do this. Mission command, the Army's preferred command and control method, directs the application of full spectrum operations to seize, retain, and exploit the initiative and achieve decisive results.

THE OPERATIONAL CONCEPT

3-1. The Army's operational concept is the core of its doctrine. It must be uniformly known and understood throughout the Service. The operational concept frames how Army forces, operating as part of a joint force, conduct operations. It describes how Army forces adapt to meet the distinct requirements of land operations. The concept is broad enough to describe operations now and in the near future. It is flexible enough to apply in any situation worldwide.

3-2. The Army's operational concept is *full spectrum operations:* **Army forces combine offensive, defensive, and stability or civil support operations simultaneously as part of an interdependent joint force to seize, retain, and exploit the initiative, accepting prudent risk to create opportunities to achieve decisive results. They employ synchronized action—lethal and nonlethal—proportional to the mission and informed by a thorough understanding of all variables of the operational environment. Mission command that conveys intent and an appreciation of all aspects of the situation guides the adaptive use of Army forces.** (See figure 3-1.)

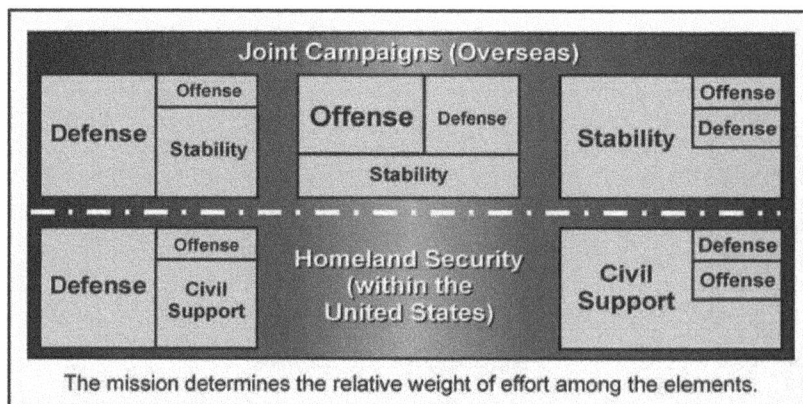

Joint Campaigns (Overseas)

| Defense | Offense / Stability | Offense | Defense / Stability | Stability | Offense / Defense |

Homeland Security (within the United States)

| Defense | Offense / Civil Support | | Civil Support | Defense / Offense |

The mission determines the relative weight of effort among the elements.

Figure 3-1. Full spectrum operations—the Army's operational concept

3-3. Full spectrum operations require continuous, simultaneous combinations of offensive, defensive, and stability or civil support tasks. In all operations, commanders seek to seize, retain, and exploit the initiative while synchronizing their actions to achieve the best effects possible. Operations conducted outside the United States and its territories simultaneously combine three elements—offense, defense, and stability. Within the United States and its territories, operations combine the elements of civil support, defense, and offense in support of civil authority. Army forces operate using mutually supporting lethal and nonlethal capabilities.

3-4. Army forces use offensive and defensive operations to defeat the enemy on land. They simultaneously execute stability or civil support operations to interact with the populace and civil authorities. In most domestic operations, Army forces perform only civil support tasks. However, an extreme emergency, such as an attack by a hostile foreign power, may require simultaneous combinations of offensive, defensive, and civil support tasks. Stability tasks typically dominate peace operations, peacetime military engagement, and some limited interventions. For example, foreign humanitarian assistance operations involve primarily stability tasks with minor defensive tasks and no offensive element.

3-5. Full spectrum operations begin with the commander's concept of operations. This single, unifying idea provides direction for the entire operation. Based on a specific idea of how to accomplish the mission, the concept of operations is refined during planning. Commanders adjust it throughout the operation as subordinates develop the situation or conditions change. Often, subordinates acting on the higher commander's intent develop the situation in ways that exploit unforeseen opportunities. Mission command is therefore an essential complement to full spectrum operations.

3-6. Operations today require versatile, well-trained units and tough, adaptive commanders. There is no set formula for applying landpower. Each campaign and major operation requires an original design and flexible execution. Army forces must be able to operate as part of a joint or multinational force anywhere on the spectrum of conflict. Varying combinations of the elements of full spectrum operations appropriate to the situation are necessary as well. The concept of operations combines and weights these elements as the situation requires.

3-7. Full spectrum operations involve continuous interaction between friendly forces and multiple groups in the operational area. In addition to enemy forces and the local populace, Soldiers deal with multinational partners, adversaries, civil authorities, business leaders, and other civilian agencies. This interaction is simple in concept but complex in application. For example, enemies and adversaries may consist of multiple competing elements. Civil authorities range from strategic-level leaders to local government officials to religious leaders. Populations may include people of differing tribes, ethnic groups, and nationalities. Within the United States and its territories, the roles and responsibilities of Army forces and civil authorities are substantially different from overseas. For that reason, Army forces conduct civil support operations domestically and stability operations overseas, even though stability and civil support operations have many similarities.

3-8. The operational concept addresses more than combat between armed opponents. Army forces conduct operations in the midst of populations. This requires Army forces to defeat the enemy and simultaneously shape civil conditions. Offensive and defensive tasks defeat enemy forces; stability tasks shape civil conditions. Winning battles and engagements is important but alone may not be decisive. Shaping civil conditions (in concert with civilian organizations, civil authorities, and multinational forces) is just as important to campaign success. In many joint operations, stability or civil support are often more important than the offense and defense.

3-9. The emphasis on different elements of full spectrum operations changes with echelon, time, and location. In an operation dominated by stability, part of the force might be conducting simultaneous offensive and defensive tasks. Within the United States, civil support operations may be the only activity actually conducted. In short, no single element is always more important than the others. Rather, *simultaneous combinations* of the elements, which commanders constantly adapt to conditions, are the key to successful land operations.

3-10. Managing training for full spectrum operations presents challenges for leaders at all echelons. Training for offensive and defensive tasks develops discipline, endurance, unit cohesion, and tolerance for uncertainty. It prepares Soldiers and units to address the ambiguities and complexities inherent in stability and civil support operations as well. However, operational experience demonstrates that forces trained exclusively for offensive and defensive tasks are not as proficient at stability tasks as those trained specifically for stability. For maximum effectiveness, stability and civil support tasks require dedicated training, similar to training for offensive and defensive tasks. Likewise, forces involved in protracted stability or civil support operations require intensive training to regain proficiency in offensive or defensive tasks before engaging in large-scale combat operations. Effective training reflects a balance among the elements of full spectrum operations that produces and sustains proficiency in all of them. Commanders adjust their emphasis by developing their core mission-essential task list. (FM 7-0 discusses training management.)

INITIATIVE

3-11. All Army operations aim to seize, retain, and exploit the initiative and achieve decisive results. *Operational initiative* **is setting or dictating the terms of action throughout an operation.** Initiative gives all operations the spirit, if not the form, of the offense. It originates in the principle of the offensive. The principle of the offensive is not just about attacking. It is about seizing, retaining, and exploiting the initiative as the surest way to achieve decisive results. It requires positive action to change both information and the situation on the ground. Risk and opportunity are intrinsic in seizing the initiative. To seize the initiative, commanders evaluate and accept prudent risks. Opportunities never last long. Unless commanders are willing to accept risk and then act, the adversary is likely to close the window of opportunity and exploit friendly inaction. Once they seize the initiative, Army forces exploit the opportunities it creates. Initiative requires constant effort to control tempo while maintaining freedom of action. The offensive mindset, with its focus on initiative, is central to the Army's operational concept and guides all leaders in the performance of their duty. It emphasizes opportunity created by action through full spectrum operations, whether offensive, defensive, stability, or civil support.

3-12. In combat operations, commanders force the enemy to respond to friendly action. In the offense, it is about taking the fight to the enemy and never allowing enemy forces to recover from the initial shock of the attack. In the defense, it is about preventing the enemy from achieving success and then counterattacking to seize the initiative. The object is not just to kill enemy personnel and destroy their equipment. It is to compel the enemy to react continuously and finally to be driven into untenable positions. Retaining the initiative pressures enemy commanders into abandoning their preferred options, accepting too much risk, or making costly mistakes. As enemy mistakes occur, friendly forces seize opportunities and create new avenues for exploitation. The ultimate goal is to break the enemy's will through relentless pressure.

3-13. In stability and civil support operations, initiative is about improving civil conditions and applying combat power to prevent the situation from deteriorating. Commanders identify objectives that may be nonmilitary but are critical to achieving the end state. Such objectives may include efforts to ensure effective governance, reconstruction projects that promote social well-being, and consistent actions to improve public safety. All of these contribute to retaining and exploiting the initiative in stability operations. An enemy insurgent, for example, cannot allow stability efforts to succeed without serious consequences and must react. As the enemy reacts, Army forces maintain the initiative by modifying their own lethal and nonlethal actions, forcing the enemy to change plans and remain on the defensive. Army forces retain the initiative by anticipating both enemy actions and civil requirements and by acting positively to address them. In civil support, Army forces take action to restore normalcy. Soldiers work closely with their civilian counterparts to remedy the conditions threatening lives, property, and domestic order. In some situations, rapid and determined action by Army forces becomes the stimulus for a demoralized civilian community to begin recovery. In every situation, commanders complement their actions with information engagement. (Paragraphs 7-10 through 7-22 discuss information engagement.)

3-14. Seizing, retaining, and exploiting the initiative depends on *individual initiative*—**the willingness to act in the absence of orders, when existing orders no longer fit the situation, or when unforeseen opportunities or threats arise.** Military history contains many instances where a subordinate's action or inaction significantly affected the tactical, operational, or even strategic situation. When opportunity occurs, it is often fleeting. Subordinate leaders need to act quickly, even as they report the situation to their commanders. Individual initiative is the key component of mission command. Full spectrum operations depend on subordinate commanders exercising individual initiative and higher commanders giving them the authority to do so. (See paragraphs 3-29 through 3-35.)

SIMULTANEITY AND SYNCHRONIZATION

3-15. Simultaneously executing the elements of full spectrum operations requires the synchronized application of combat power. *Synchronization* is the arrangement of military actions in time, space, and purpose to produce maximum relative combat power at a decisive place and time (JP 2-0). It is the ability to execute multiple related and mutually supporting tasks in different locations at the same time, producing greater effects than executing each in isolation. Synchronization is a means, not an end. Commanders balance it with agility and initiative; they never surrender the initiative for the sake of synchronization. Rather, they synchronize activities to best facilitate mission accomplishment. Excessive synchronization can lead to over-control, which limits the initiative of subordinates.

3-16. Simultaneity means doing multiple things at the same time. It requires the ability to conduct operations in depth and to orchestrate them so that their timing multiplies their effectiveness. Commanders consider their entire area of operations the enemy, the information tasks necessary to shape the operational environment, and civil conditions. Then they mount simultaneous operations that immobilize, suppress, or shock the enemy. Such actions nullify the enemy's ability to conduct synchronized, mutually supporting actions. Army forces increase the depth of their operations through combined arms, advanced information systems, and joint capabilities. Because Army forces conduct operations across large areas, the enemy faces many potential friendly actions. Executing operations in depth is equally important in stability operations; commanders act to keep threats from operating outside the reach of friendly forces. In civil support operations and some stability operations, depth includes conducting operations that reach all citizens in the area of operations, bringing relief as well as hope. (Paragraphs 6-85 through 6-87 discuss depth at the operational level.)

LETHAL AND NONLETHAL ACTIONS

3-17. An inherent, complementary relationship exists between using lethal force and applying military capabilities for nonlethal purposes. Though each situation requires a different mix of violence and restraint, lethal and nonlethal actions used together complement each other and create dilemmas for opponents. Lethal actions are critical to accomplishing offensive and defensive missions. However, nonlethal actions are also important contributors to full spectrum operations, regardless of which element dominates. Finding ways to accomplish the mission with an appropriate mix of lethal and nonlethal force has become an important consideration for every commander. Commanders analyze the situation carefully to achieve a balance between lethal and nonlethal actions.

Lethal Actions

3-18. Offensive and defensive operations place a premium on employing the lethal effects of combat power against the enemy. In these operations, speed, surprise, and shock are vital considerations. Historically, the side better able to combine them defeats its opponent rapidly while incurring fewer losses. Such victories create opportunities for exploitation. In some operations, the effects of speed, surprise, and shock are enough to collapse organized resistance. Such a collapse occurred in the offensive phase of Operation Iraqi Freedom in 2003.

3-19. Speed is swiftness of action. It allows a force to act before the enemy is ready or before the situation deteriorates further. Speed requires being able to adjust operations quickly to dynamic conditions. It increases opportunities to exploit momentary tactical advantages and expand them to retain and exploit the initiative. Wherever possible, Army forces exploit their advantages in command and control, tactical mobility, and joint capabilities to operate at a higher tempo than the enemy. Delegating decisionmaking authority to subordinates through mission command allows commanders to increase the speed of decisionmaking and execution in most situations. Rapid friendly action may surprise the enemy and create opportunities.

3-20. Surprise is achieved by acting at a time, acting in a place, or using methods to which the enemy cannot effectively react or does not expect. Speed contributes to surprise. So does executing operations simultaneously and in depth. Exploiting advantages gained through security, military deception, and aggressive intelligence, surveillance, and reconnaissance (ISR) operations contributes to surprise. Surprise is essential when executing offensive missions that complement stability operations (such as a raid). It mitigates the effectiveness of enemy early warning networks within the local populace.

3-21. Shock results from applying overwhelming violence. Combat power applied with enough speed and magnitude to overwhelm the enemy produces it. Shock slows and disrupts enemy operations. It is usually transient, but while it lasts, shock may paralyze the enemy's ability to fight. Sometimes the psychological effects of threatening to use overwhelming violence can also produce shock. Shock is often greater when generated with asymmetric means. Joint forces create opportunities to increase it by using capabilities against which the enemy has limited defense. Surprise and speed magnify the effects of shock.

Nonlethal Actions

3-22. Army forces employ a variety of nonlethal means. Stability operations often involve using military capabilities to perform such tasks as restoring essential services. Civil support operations are characterized by providing constructive support to civil authorities. However, demonstrating the potential for lethal action (by actions such as increased military presence in an area) often contributes to maintaining order during stability and some civil support operations. Other examples include such actions as pre-assault warnings and payments for collateral damage. Some nonlethal actions, such as information engagement, are common to all operations.

3-23. Friendly and enemy forces continuously struggle for information advantages while conducting operations in the physical domains. Friendly information actions shape the operational environment by attacking the enemy's command and control system, defending against electronic attacks, and protecting friendly information. Command and control warfare contributes to the success of offensive and defensive operations. (See paragraphs 7-23 through 7-30.) It employs many nonlethal means to attack the enemy's command and control capabilities. Increasingly sophisticated capabilities allow Army forces to identify, disrupt, and exploit enemy communications (including networks). These actions may keep the enemy from massing combat power effectively or synchronizing combined arms operations. Commanders may use electromagnetic means alone or with maneuver and lethal fires.

3-24. Nonlethal actions in combat include a wide range of intelligence-gathering, disruptive, and other activities. Effective maneuver and fires require timely, accurate intelligence and an effective command and control system. The threat of detection often compels the enemy to limit or cease operations. This inaction allows friendly forces to seize the initiative. Interference with enemy command and control through nonlethal means can also limit enemy effectiveness and increase its exposure to attack.

3-25. The United States continues to develop nonlethal weapons that allow commanders to apply force without killing or crippling an enemy. These weapons provide options in situations that restrict the use of lethal force or when enemy fighters intermix with noncombatants. Furthermore, nonlethal means can mitigate the indirect effects on noncombatants of lethal actions directed against the enemy.

3-26. Stability and civil support operations emphasize nonlethal, constructive actions by Soldiers working among noncombatants. Civil affairs personnel have a major role. In stability operations, they work with and through host-nation agencies and other civilian organizations to enhance the host-nation government's legitimacy. Commanders use continuous information engagement shaped by intelligence to inform, influence, and persuade the local populace within limits prescribed by U.S. law. They also integrate information engagement with stability tasks to counter false and distorted information and propaganda.

3-27. Nonlethal, constructive actions can persuade the local populace to withhold support from the enemy and provide information to friendly forces. Loss of popular support presents the enemy with two bad choices: stay and risk capture or depart and risk exposure to lethal actions in less populated areas. Commanders focus on managing the local populace's expectations and countering rumors. However, they recognize that their Soldiers' actions, positive and negative, are the major factor in the populace's perception of Army forces.

3-28. The moral advantage provided by the presence of well-trained, well-equipped, and well-led forces can be a potent nonlethal capability. It creates fear and doubt in the minds of the enemy and may deter adversaries. This effect is important in many stability-dominated operations. Even though stability operations emphasize nonlethal actions, the ability to engage potential enemies with decisive lethal force remains a sound deterrent. Enemy commanders may curtail their activities and avoid combat if they perceive Army forces as highly capable and willing to use precise, lethal force. This permits Army forces to extend the scope and tempo of nonlethal actions.

MISSION COMMAND AND FULL SPECTRUM OPERATIONS

3-29. The Army's preferred method of exercising command and control is mission command. ***Mission command* is the conduct of military operations through decentralized execution based on mission orders. Successful mission command demands that subordinate leaders at all echelons exercise disciplined initiative, acting aggressively and independently to accomplish the mission within the commander's intent**. Mission command gives subordinates the greatest possible freedom of action. Commanders focus their orders on the purpose of the operation rather than on the details of how to perform assigned tasks. They delegate most decisions to subordinates. This minimizes detailed control and empowers subordinates' initiative. Mission command emphasizes timely decisionmaking, understanding the higher commander's intent, and clearly identifying the subordinates' tasks necessary to achieve the desired end state. It improves subordinates' ability to act effectively in fluid, chaotic situations.

3-30. Mission command requires an environment of trust and mutual understanding. Respect and full familiarity with the commander's intent and concept of operations are also essential. Mission command applies to all operations across the spectrum of conflict. The elements of mission command are—

- Commander's intent.
- Subordinates' initiative.
- Mission orders, which include—
    - A brief concept of operations.
    - Minimum necessary control measures.
- Resource allocation.

3-31. Chaos and uncertainty dominate land conflict. Predictability is rare, making centralized decisionmaking and orderly processes ineffective. In addition, commanders must contend with a thinking, adaptive enemy. Under such conditions, leaders of forces in contact can often see and act on immediate opportunities and threats better than their superiors can. Delegating the greatest possible authority to subordinates helps the force adapt the operation to the situation quickly and retain the initiative.

3-32. While mission command restrains higher level commanders from micromanaging subordinates, it does not remove them from the fight. Rather, mission command frees these commanders to focus on accomplishing their higher commander's intent and on critical decisions only they can make. Higher commanders anticipate developments, allocate resources to exploit successes, and intervene to shape the operation as necessary. Mission command is ideally suited to an environment of complexity and uncertainty.

3-33. Mission command tends to be decentralized, informal, and flexible. Orders and plans are as brief and simple as possible. The fundamental basis of mission command is trust and mutual understanding between superiors and subordinates, an atmosphere that the senior commander must cultivate. Mission command counters the uncertainty of war by empowering subordinates at the scene to make decisions quickly. Commanders rely on their subordinates' ability to coordinate with one another, using the human capacity to understand with minimum verbal information exchange.

3-34. In any operation, the situation may change rapidly. The speed and violence of modern operations add to combat's inherent chaos and disorder. Operations the commander envisioned may bear little resemblance to actual events. Even with the most advanced information systems, higher headquarters often have difficulty understanding the situation on the ground. Perhaps most important, opportunities for exploitation or counterattack may occur suddenly. Subordinate commanders need maximum latitude to take advantage of such situations and meet the higher commander's intent when the original orders no longer apply. Full spectrum operations require leaders schooled in independent decisionmaking, aggressiveness, and risk taking in an environment of mission orders and mission command at every level.

3-35. Mission command applies to all operations across the spectrum of conflict. Often the operational environment encountered during stability and civil support operations proves more complex than that encountered in offensive and defensive operations. The continuous, often volatile, interaction of brigades and smaller units with the local populace during stability operations requires leaders willing to exercise initiative. They must be able and willing to solve problems without constantly referring to higher headquarters. Mission command encourages commanders to act promptly, consistently, and decisively in all situations. Under mission command, commanders explain not only the tasks assigned and their immediate purpose but also the higher commander's intent. Doing this helps junior commanders and their Soldiers understand what is expected of them and what constraints to apply. Most importantly, they understand the mission's purpose and context. The commander's intent also guides subordinates working with agencies not under military control.

THE ELEMENTS OF FULL SPECTRUM OPERATIONS

3-36. Full spectrum operations require simultaneous combinations of four elements—offense, defense, and stability or civil support. Figure 3-2 lists the elements of full spectrum operations, the primary tasks associated with them, and the purposes of each element. Each primary task has numerous associated subordinate tasks. When combined with who (unit), when (time), where (location), and why (purpose), the primary tasks become mission statements.

Offensive Operations	Defensive Operations
Primary Tasks • Movement to contact • Attack • Exploitation • Pursuit **Purposes** • Dislocate, isolate, disrupt, and destroy enemy forces • Seize key terrain • Deprive the enemy of resources • Develop intelligence • Deceive and divert the enemy • Create a secure environment for stability operations	**Primary Tasks** • Mobile defense • Area defense • Retrograde **Purposes** • Deter or defeat enemy offensive operations • Gain time • Achieve economy of force • Retain key terrain • Protect the populace, critical assets, and infrastructure • Develop intelligence
Stability Operations	Civil Support Operations
Primary Tasks • Civil security • Civil control • Restore essential services • Support to governance • Support to economic and infrastructure development **Purposes** • Provide a secure environment • Secure land areas • Meet the critical needs of the populace • Gain support for host-nation government • Shape the environment for interagency and host-nation success	**Primary Tasks** • Provide support in response to disaster or terrorist attack • Support civil law enforcement • Provide other support as required **Purposes** • Save lives • Restore essential services • Maintain or restore law and order • Protect infrastructure and property • Maintain or restore local government • Shape the environment for interagency success

Figure 3-2. The elements of full spectrum operations

OFFENSIVE OPERATIONS

3-37. *Offensive operations* **are combat operations conducted to defeat and destroy enemy forces and seize terrain, resources, and population centers. They impose the commander's will on the enemy.** In combat operations, the offense is the decisive element of full spectrum operations. Against a capable, adaptive enemy, the offense is the most direct and sure means of seizing, retaining, and exploiting the initiative to achieve decisive results. Executing offensive operations compels the enemy to react, creating or revealing weaknesses that the attacking force can exploit. Successful offensive operations place tremendous pressure on defenders, creating a cycle of deterioration that can lead to their disintegration. This was the case in early 2003 in Iraq, when coalition operations led to the collapse of the Iraqi military and ultimately the Baathist regime of Saddam Hussein.

3-38. While strategic, operational, or tactical considerations may require defending, defeating an enemy at any level sooner or later requires shifting to the offense. Even in the defense, seizing and retaining the initiative requires executing offensive operations at some point. The more fluid the battle, the more true this is.

3-39. Effective offensive operations capitalize on accurate intelligence regarding the enemy, terrain and weather, and civil considerations. Commanders maneuver their forces to advantageous positions before making contact. However, commanders may shape conditions by deliberately making contact to develop the situation and mislead the enemy. In the offense, the decisive operation is a sudden, shattering action against an enemy weakness that capitalizes on speed, surprise, and shock. If that operation does not destroy the enemy, operations continue until enemy forces disintegrate or retreat to where they no longer pose a threat. (Paragraphs 5-59 through 5-64 discuss decisive, shaping, and sustaining operations.)

Primary Offensive Tasks

3-40. At the operational level, offensive operations defeat enemy forces that control important areas or contest the host-nation government's authority. The joint force conducts operations throughout its operational area. Army forces attack using ground and air maneuver to achieve objectives that conclude the campaign or move it to a subsequent phase. In expeditionary campaigns and major operations, operational maneuver includes deploying land forces to positions that facilitate joint force offensive action. Operational-level offensives in counterinsurgency may be conducted to eliminate insurgent sanctuaries. Counterinsurgencies usually combine offensive and stability tasks to achieve decisive results.

3-41. In offensive operations, a force often transitions from one offensive task to another without pausing. For example, an attack can lead to exploitation and then pursuit, or to exploitation followed by another attack as enemy forces rally. Army forces perform the following primary offensive tasks. (FM 3-90 discusses them in detail.)

Movement to Contact

3-42. A movement to contact develops the situation and establishes or regains contact. It also creates favorable conditions for subsequent tactical actions. Forces executing this task seek to make contact with the smallest friendly force feasible. On contact, the commander has five options: attack, defend, bypass, delay, or withdraw. Movements to contact include search and attack and cordon and search operations.

Attack

3-43. An attack destroys or defeats enemy forces, seizes and secures terrain, or both. Attacks require maneuver supported by direct and indirect fires. They may be either decisive or shaping operations. Attacks may be hasty or deliberate, depending on the time available for planning and preparation. Commanders execute hasty attacks when the situation calls for immediate action with available forces and minimal preparation. They conduct deliberate attacks when they have more time to plan and prepare. Success depends on skillfully massing the effects of all the elements of combat power.

Exploitation

3-44. An exploitation rapidly follows a successful attack and disorganizes the enemy in depth. Exploitations seek to expand an attack to the point where enemy forces have no alternatives but to surrender or flee. Commanders of exploiting forces receive the greatest possible latitude to accomplish their missions. They act with great aggressiveness, initiative, and boldness. Exploitations may be local or major. Local exploitations take advantage of tactical opportunities, foreseen or unforeseen. Division and higher headquarters normally conduct major exploitations using mobile forces to transform tactical success into a pursuit.

Pursuit

3-45. A pursuit is designed to catch or cut off a hostile force attempting to escape with the aim of destroying it. Pursuits often follow successful exploitations. However, they can develop at any point when enemy forces are beginning to disintegrate or disengage. Pursuits occur when the enemy fails to organize a defense and attempts to disengage. If it becomes apparent that enemy resistance has broken down entirely and enemy forces are fleeing, a force can transition to a pursuit from any type of offensive or defensive operation. Pursuits require speed and decentralized control.

Purposes of Offensive Operations

3-46. Seizing, retaining, and exploiting the initiative is the essence of the offense. Offensive operations seek to throw enemy forces off balance, overwhelm their capabilities, disrupt their defenses, and ensure their defeat or destruction by synchronizing and applying all the elements of combat power. The offensive operation ends when it destroys or defeats the enemy, reaches a limit of advance, or approaches culmination. Army forces conclude an offense in one of four ways: consolidating gains through stability operations, resuming the attack, transitioning to the defense, or preparing for future operations. Army forces conduct offensive operations for the following purposes. (FM 3-90 discusses these purposes in detail.)

Dislocate, Isolate, Disrupt, and Destroy Enemy Forces

3-47. Well-executed offensive operations dislocate, isolate, disrupt, and destroy enemy forces. If destruction is not feasible, offensive operations compel enemy forces to retreat. Offensive maneuver seeks to place the enemy at a positional disadvantage. This allows friendly forces to mass overwhelming effects while defeating parts of the enemy force in detail before the enemy can escape or be reinforced. When required, friendly forces close with and destroy the enemy in close combat. Ultimately, the enemy surrenders, retreats in disorder, or is eliminated altogether.

Seize Key Terrain

3-48. Offensive maneuver may seize terrain that provides the attacker with a decisive advantage. The enemy either retreats or risks defeat or destruction. If enemy forces retreat or attempt to retake the key terrain, they are exposed to fires and further friendly maneuver.

Deprive the Enemy of Resources

3-49. At the operational level, offensive operations may seize control of major population centers, seats of government, production facilities, and transportation infrastructure. Losing these resources greatly reduces the enemy's ability to resist. In some cases, Army forces secure population centers or infrastructure and prevent irregular forces from using them as a base or benefitting from the resources that they generate.

Develop Intelligence

3-50. Enemy deception, concealment, and security may prevent friendly forces from gaining necessary intelligence. Some offensive operations are conducted to develop the situation and discover the enemy's intent, disposition, and capabilities.

Deceive and Divert the Enemy

3-51. Offensive operations distract enemy intelligence, surveillance, and reconnaissance. They may cause the enemy to shift reserves away from the friendly decisive operation.

Create a Secure Environment for Stability Operations

3-52. Stability operations cannot occur if significant enemy forces directly threaten or attack the local populace. Offensive operations destroy or isolate the enemy so stability operations can proceed. Offensive operations against insurgents help keep them off balance. These actions may force insurgents to defend their bases, thus keeping them from attacking.

DEFENSIVE OPERATIONS

3-53. *Defensive operations* **are combat operations conducted to defeat an enemy attack, gain time, economize forces, and develop conditions favorable for offensive or stability operations**. The defense alone normally cannot achieve a decision. However, it can create conditions for a counteroffensive operation that lets Army forces regain the initiative. Defensive operations can also establish a shield behind which stability operations can progress. Defensive operations counter enemy offensive operations. They defeat attacks, destroying as much of the attacking enemy as possible. They also preserve control over land, resources, and populations. Defensive operations retain terrain, guard populations, and protect critical capabilities against enemy attacks. They can be used to gain time and economize forces so offensive tasks can be executed elsewhere.

3-54. Successful defensive operations share the following characteristics: preparation, security, disruption, massed effects, and flexibility. Successful defenses are aggressive. Commanders use all available means to disrupt enemy forces. They disrupt attackers and isolate them from mutual support to defeat them in detail. Isolation includes extensive use of command and control warfare. Defenders seek to increase their freedom of maneuver while denying it to attackers. Defending commanders use every opportunity to transition to the offense, even if temporarily. As attackers' losses increase, they falter and the initiative shifts to the defenders. These situations are favorable for counterattacks. Counterattack opportunities rarely last long; defenders strike swiftly when the attackers culminate. Surprise and speed complement shock and allow counterattacking forces to seize the initiative and overwhelm the attackers.

3-55. Conditions may not support immediate offensive operations during force projection. In those cases, initial-entry forces defend while the joint force builds combat power. Initial-entry forces should include enough combat power to deter, attack, or defend successfully.

Primary Defensive Tasks

3-56. At the operational level, an enemy offensive may compel joint forces to conduct major defensive operations. Such operations may require defeating or preventing attacks across international borders, defeating conventional attacks, or halting an insurgent movement's mobilization. Operational defenses may be executed anywhere in the operational area. The following primary tasks are associated with the defense. Defending commanders combine these tasks to fit the situation. (FM 3-90 discusses them in detail.)

Mobile Defense

3-57. In a mobile defense, the defender withholds a large portion of available forces for use as a striking force in a counterattack. Mobile defenses require enough depth to let enemy forces advance into a position that exposes them to counterattack. The defense separates attacking forces from their support and disrupts the enemy's command and control. As enemy forces extend themselves in the defended area and lose momentum and organization, the defender surprises and overwhelms them with a powerful counterattack.

Area Defense

3-58. In an area defense, the defender concentrates on denying enemy forces access to designated terrain for a specific time, limiting their freedom of maneuver and channeling them into killing areas. The defender retains terrain that the attacker must control in order to advance. The enemy force is drawn into a series of kill zones where it is attacked from mutually supporting positions and destroyed, largely by fires. Most of the defending force is committed to defending positions while the rest is kept in reserve. Commanders use the reserve to preserve the integrity of the defense through reinforcement or counterattack.

Retrograde

3-59. Retrograde involves organized movement away from the enemy. This includes delays, withdrawals, and retirements. Retrograde operations gain time, preserve forces, place the enemy in unfavorable positions, or avoid combat under undesirable conditions.

Mobile and Static Elements in the Defense

3-60. All three primary defensive tasks use mobile and static elements. In mobile defenses, static positions help control the depth and breadth of the enemy penetration and retain ground from which to launch counterattacks. In area defenses, commanders closely integrate mobile patrols, security forces, sensors, and reserves to cover gaps among defensive positions. In retrograde operations, some units conduct area or mobile defenses along with security operations to protect other units executing carefully controlled maneuver or movement rearward. Static elements fix, disrupt, turn, or block the attackers and gain time for other forces to pull back. Mobile elements maneuver constantly to confuse the enemy and prevent enemy exploitation.

Purposes of Defense Operations

3-61. Defending forces await the enemy's attack and counter it. Waiting for the attack is not a passive activity. Commanders conduct aggressive security operations and intelligence, surveillance, and reconnaissance. Such actions locate enemy forces and deny them information. Defenders engage enemy forces with fires, spoiling attacks, and security operations to weaken them before they reach the main battle area. Commanders use combined arms and joint capabilities to attack enemy vulnerabilities and seize the initiative. Army forces conduct defensive operations for the following purposes. (FM 3-90 discusses these purposes in detail.)

Deter or Defeat Enemy Offensive Operations

3-62. The primary purpose of the defense is to deter or defeat enemy offensive operations. Successful defenses stall enemy actions and create opportunities to seize the initiative. Defensive operations may deter potential aggressors if they believe that breaking the friendly defense would be too costly.

Gain Time

3-63. Commanders may conduct a defense to gain time. Such defensive operations succeed by slowing or halting an attack while allowing friendly reserves enough time to reinforce the defense. Delaying actions trade space for time to improve defenses, expose enemy forces to joint attack, and prepare counterattacks.

Achieve Economy of Force

3-64. The defense is also used to achieve economy of force. Astute use of terrain, depth, and security operations allows friendly forces to minimize resources used defensively. This allows commanders to concentrate combat power for offensive operations.

Retain Key Terrain

3-65. The mission of many defensive operations is to retain key terrain. Such defenses are necessary to prevent enemy forces from occupying key terrain. Control of key terrain can sway the outcome of the battle or engagement depending on which side controls it. In operations dominated by stability tasks, friendly bases become key terrain.

Protect the Populace, Critical Assets, and Infrastructure

3-66. Defense of the local populace and vital assets supports stability operations and allows Army forces to receive greater support from the host nation. Army forces protect military and civilian areas that are important to success and provide indirect support to operations worldwide. Achieving this purpose begins with defenses around lodgments and bases, ensuring freedom of action. This protection is very important in counterinsurgency operations where some facilities have significant economic and political value as opposed to tactical military importance.

Develop Intelligence

3-67. As with the offense, defensive operations may develop intelligence. The more successful the defense, the more Army forces learn about the enemy. A particular phase or task within a defense (for example, a covering force mission) may be conducted to satisfy commander's critical information requirements about the enemy's direction of attack and main effort.

STABILITY OPERATIONS

3-68. *Stability operations* encompass various military missions, tasks, and activities conducted outside the United States in coordination with other instruments of national power to maintain or reestablish a safe and secure environment, provide essential governmental services, emergency infrastructure reconstruction, and humanitarian relief (JP 3-0). Stability operations can be conducted in support of a host-nation or interim government or as part of an occupation when no government exists. Stability operations involve both coercive and constructive military actions. They help to establish a safe and secure environment and facilitate reconciliation among local or regional adversaries. Stability operations can also help establish political, legal, social, and economic institutions and support the transition to legitimate local governance. Stability operations must maintain the initiative by pursing objectives that resolve the causes of instability. (See paragraph 3-13.) Stability operations cannot succeed if they only react to enemy initiatives.

3-69. Coordination, integration, and synchronization between host-nation elements, other government agencies, and Army forces are enhanced by transparency and credibility. The degree to which the host nation cooperates is fundamental. Commanders publicize their mandate and intentions. Within the limits of operations security, they make the populace aware of the techniques used to provide security and control. Actions on the ground reinforced by a clear and consistent message produce transparency. This transparency reinforces credibility. Credibility reflects the populace's assessment of whether the force can accomplish the mission. Army forces require the structure, resources, and rules of engagement appropriate to accomplishing the mission and discharging their duties swiftly and firmly. They must leave no doubt as to their capability and intentions.

3-70. Civil affairs activities enhance the relationship between military forces and civil authorities in areas with military forces. They involve applying civil affairs functional specialty skills to areas normally under the responsibility of civil government. These operations involve establishing, maintaining, influencing, or exploiting relations between military forces and all levels of host-nation government agencies. These activities are fundamental to executing stability tasks. Civil affairs personnel, other Army forces, other government agencies, or a combination of all three perform these tasks.

3-71. Civil affairs units and personnel develop detailed civil considerations assessments. These include information about infrastructure, civilian institutions, and the attitudes and activities of civilian leaders, populations, and organizations. These assessments may reveal that a viable host-nation government does not exist or is incapable of performing its functions. In such cases, Army forces may support or exercise governmental authority until a host-nation civil authority is established. (JPs 3-57 and 3-57.1 and FMs 3-05.40 and 3-05.401 contain civil affairs doctrine.)

Primary Stability Tasks

3-72. The combination of tasks conducted during stability operations depends on the situation. In some operations, the host nation can meet most or all of the population's requirements. In those cases, Army forces work with and through host-nation authorities. Commanders use civil affairs activities to mitigate how the military presence affects the populace and vice versa. Conversely, Army forces operating in a failed state may need to support the well-being of the local populace. That situation requires Army forces to work with civilian agencies to restore basic capabilities. Again, civil affairs activities are important in establishing the trust between Army forces and civilian organizations required for effective, working relationships.

3-73. Stability operations may be necessary to develop host-nation capacities for security and control of security forces, a viable market economy, the rule of law, and an effective government. Army forces develop these capabilities by working with the host nation. The goal is a stable civil situation sustainable by host-nation assets without Army forces. Security, the health of the local economy, and the capability of self-government are related. Without security, the local economy falters. A functioning economy provides employment and reduces the dependence of the population on the military for necessities. Security and economic stability precede an effective and stable government.

3-74. Stability operations require the absence of major threats to friendly forces and the populace. As offensive operations clear areas of hostile forces, part of the force secures critical infrastructure and populated areas. Establishing civil security and essential services are implied tasks for commanders during any combat operation. Commanders should act to minimize and relieve civilian suffering. However, if a unit is decisively engaged in combat operations, it should not be diverted from its mission to perform stability tasks.

3-75. Commanders plan to minimize the effects of combat on the populace. They promptly inform their higher headquarters of civilian requirements and conditions that require attention. As civil security is established, the force returns territory to civil authorities' control when feasible. Transitions to civil authority require coordinating and integrating civilian and military efforts. Unified action is crucial. Properly focused, effectively executed stability tasks prevent population centers from degenerating into civil unrest and becoming recruiting areas for opposition movements or insurgencies.

3-76. Army forces perform five primary stability tasks. (When revised, FM 3-07 will discuss these tasks in detail.)

Civil Security

3-77. Civil security involves protecting the populace from external and internal threats. Ideally, Army forces defeat external threats posed by enemy forces that can attack population centers. Simultaneously, they assist host-nation police and security elements as the host nation maintains internal security against terrorists, criminals, and small, hostile groups. In some situations, no adequate host-nation capability for civil security exists. Then, Army forces provide most civil security while developing host-nation capabilities. For the other stability tasks to be effective, civil security is required. As soon the host-nation security forces can safely perform this task, Army forces transition civil security responsibilities to them.

Civil Control

3-78. Civil control regulates selected behavior and activities of individuals and groups. This control reduces risk to individuals or groups and promotes security. Civil control channels the populace's activities to allow provision of security and essential services while coexisting with a military force conducting operations. A curfew is an example of civil control.

Restore Essential Services

3-79. Army forces establish or restore the most basic services and protect them until a civil authority or the host nation can provide them. Normally, Army forces support civilian and host-nation agencies. When the host nation cannot perform its role, Army forces may provide the basics directly. Essential services include the following:

- Providing emergency medical care and rescue.
- Preventing epidemic disease.
- Providing food and water.
- Providing emergency shelter.
- Providing basic sanitation (sewage and garbage disposal).

Support to Governance

3-80. Stability operations establish conditions that enable actions by civilian and host-nation agencies to succeed. By establishing security and control, stability operations provide a foundation for transitioning authority to civilian agencies and eventually to the host nation. Once this transition is complete, commanders focus on transferring control to a legitimate civil authority according to the desired end state. Support to governance includes the following:

- Developing and supporting host-nation control of public activities, the rule of law, and civil administration.
- Maintaining security, control, and essential services through host-nation agencies. This includes training and equipping host-nation security forces and police.
- Supporting host-nation efforts to normalize the succession of power (elections and appointment of officials).

Support to Economic and Infrastructure Development

3-81. Support to economic and infrastructure development helps a host nation develop capability and capacity in these areas. It may involve direct and indirect military assistance to local, regional, and national entities.

Purposes of Stability Operations

3-82. Although Army forces focus on achieving the military end state, they ultimately need to create conditions where the other instruments of national power are preeminent. Stability operations focus on creating those conditions. The following paragraphs discuss the purposes of stability operations. (When revised, FM 3-07 will discuss these purposes in detail.)

Provide a Secure Environment

3-83. A key stability task is providing a safe, secure environment. This involves isolating enemy fighters from the local populace and protecting the population. By providing security and helping host-nation authorities control civilians, Army forces begin the process of separating the enemy from the general population. Information engagement complements physical isolation by persuading the populace to support an acceptable, legitimate host-nation government. This isolates the enemy politically and economically.

Secure Land Areas

3-84. Effective stability operations, together with host-nation capabilities, help secure land areas. Areas of population unrest often divert forces that may be urgently needed elsewhere. In contrast, stable areas may support bases and infrastructure for friendly forces, allowing commitment of forces elsewhere.

Meet the Critical Needs of the Populace

3-85. Often, stability operations are required to meet the critical needs of the populace. Army forces can provide essential services until the host-nation government or other agencies can do so.

Gain Support for Host-Nation Government

3-86. Successful stability operations ultimately depend on the legitimacy of the host-nation government—its acceptance by the populace as the governing body. All stability operations are conducted with that aim.

Shape the Environment for Interagency and Host-Nation Success

3-87. Stability operations shape the environment for interagency and host-nation success. They do this by providing the security and control necessary for host-nation and interagency elements to function, and supporting them in other key functions.

Stability Operations and Department of State Post-Conflict Technical Sectors

3-88. The stability tasks are linked to the Department of State post-conflict reconstruction and stabilization technical sectors. Normally Army forces act in support of host-nation and other civilian agencies. However, when the host nation cannot provide basic government functions, Army forces may be required to do so directly. The Department of State organizes conditions related to post-conflict stability into five sectors. (See figure 3-3.) Army stability tasks support these sectors, which are discussed below.

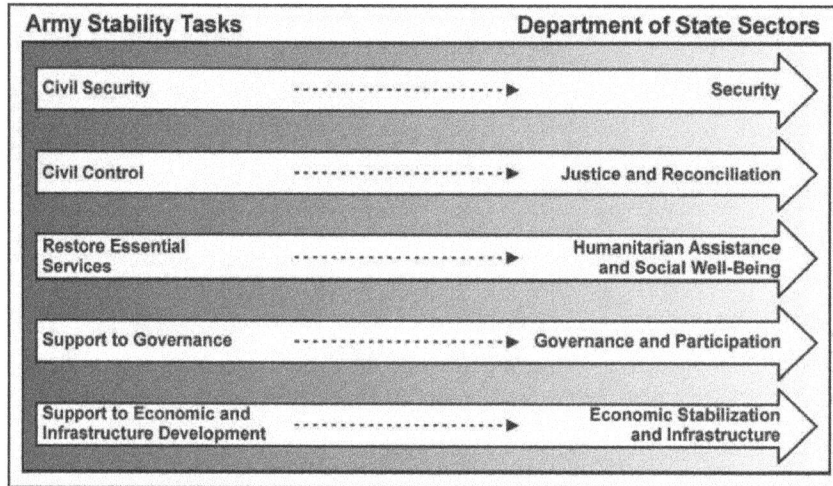

Army Stability Tasks	Department of State Sectors
Civil Security	············▶ Security
Civil Control	············▶ Justice and Reconciliation
Restore Essential Services	············▶ Humanitarian Assistance and Social Well-Being
Support to Governance	············▶ Governance and Participation
Support to Economic and Infrastructure Development	············▶ Economic Stabilization and Infrastructure

Figure 3-3. Stability tasks and Department of State technical sectors

Security

3-89. Army forces conduct operations to establish and maintain a safe and secure environment, whenever possible together with host-nation forces. They provide public order and safety and protect individuals, infrastructure, and institutions. Typically, offensive, defensive, and stability tasks all support this sector. Civil security and civil control are associated stability tasks. Initially, Army forces have the lead for this sector. Army operations should complement and be concurrent with other agencies' actions. Army forces hand over responsibility to host-nation agencies and assume a supporting role as these agencies develop the needed capabilities.

3-90. The first aim in the security sector is to limit adversaries' influence and isolate the populace from the enemy. Army forces use several methods to isolate the enemy. Isolation in some stability operations is indirect; that is, it aims to redirect, compel, and influence the populace away from supporting adversaries and toward supporting the host-nation government. Concurrently, Army forces secure the support of populations in unstable areas. If the enemy poses a significant military threat, forces engaged in stability operations defend themselves and the populace from attacks. Army forces may conduct offensive operations to physically isolate, defeat, or destroy forces that threaten the stability mission. Security is the foremost condition; it underpins all other sectors.

Justice and Reconciliation

3-91. Establishing public order and safety and providing for social reconciliation are this sector's objectives. The host nation aims to establish self-sustaining public law and order that operates according to internationally recognized standards and respects human rights and freedoms.

Humanitarian Assistance and Social Well-Being

3-92. Army forces work to reduce human suffering, disease, and privation. This sector includes programs conducted to relieve or reduce the results of conditions that present a serious threat to life or that can result in great damage to or loss of property. These conditions may be endemic or result from natural or man-made disasters.

Governance and Participation

3-93. This sector is concerned with restoring or creating effective government institutions. These efforts involve strengthening host-nation governance and rebuilding government infrastructure. They also require developing institutions that achieve sustainable peace and security, foster a sense of confidence, and support conditions for economic reconstruction. The main goal for Army forces is creating an environment conducive to stable governance. Civilian agencies are responsible for areas such as the following:

- Reestablishing the administrative framework.
- Supporting development of a national constitution.
- Supporting political reform.
- Reforming or establishing fair taxation.

Economic Stabilization and Infrastructure

3-94. Infrastructure restoration begins with meeting the basic needs of the populace. It continues by restoring economic production and distribution. The basic needs of the populace are met by reconstituting the following:

- Power.
- Transportation.
- Communications.
- Health and sanitation.
- Firefighting.
- Mortuary services.
- Environmental control.

3-95. Once the basic infrastructure is functioning, efforts shift to stabilizing the economy. Economic stabilization consists of the following:

- Restoring employment opportunities.
- Initiating market reform.
- Mobilizing domestic and foreign investment.
- Supervising monetary reform and rebuilding public structures.

Use of Force in Stability Operations

3-96. When using force, precision is as important in stability missions as applying massed, overwhelming force is in offensive and defensive operations. Commanders at every level emphasize that in stability operations, violence not precisely applied is counterproductive. Speed, surprise, and shock are vital considerations in lethal actions; perseverance, legitimacy, and restraint are vital considerations in stability and civil support operations.

3-97. The presence of armed Soldiers operating among the local populace causes tension. Discipline and strict adherence to the rules of engagement are essential but not sufficient to reassure the population. In addressing the populace's apprehension, commanders balance protecting the force, defeating enemy forces, and taking constructive action throughout the area of operations. They also stress cultural awareness in training and preparing for operations. Cultural awareness makes Soldiers more effective when operating in a foreign population and allows them to leverage local culture to enhance the effectiveness of their operations.

3-98. In peace operations, commanders emphasize impartiality in the use of force in addition to credibility and transparency. Impartiality is not neutrality. Impartiality does not imply that Army forces treat all sides equally. Force is used against threats in accordance with the rules of engagement. Fair treatment of the local populace improves the prospects for lasting peace, stability, and security.

CIVIL SUPPORT OPERATIONS

3-99. *Civil support* is Department of Defense support to U.S. civil authorities for domestic emergencies, and for designated law enforcement and other activities (JP 1-02). Civil support includes operations that address the consequences of natural or man-made disasters, accidents, terrorist attacks, and incidents in the United States and its territories. Army forces conduct civil support operations when the size and scope of events exceed the capabilities or capacities of domestic civilian agencies. Usually the Army National Guard is the first military force to respond on behalf of state authorities. In this capacity, it functions under authority of Title 32, U.S. Code, or while serving on state active duty. The National Guard is suited to conduct these missions; however, the scope and level of destruction may require states to request assistance from Federal authorities.

Primary Civil Support Tasks

3-100. Army forces perform civil support tasks under U.S. law. However, U.S. law carefully limits actions that military forces, particularly Regular Army units, can conduct in the United States and its territories. National Guard forces under state control have law enforcement authorities that Regular Army units do not have. In addition to legal differences, civil support operations always are conducted in support of state and Federal agencies. Army forces coordinate and synchronize their efforts closely with them. These agencies are trained, resourced, and equipped more extensively than similar agencies involved in stability operations overseas. In stability operations, multinational participation is typical; in civil support operations, it is the exception. Army civil support operations include three primary tasks.

Provide Support in Response to Disaster or Terrorist Attack

3-101. Policies issued by the Federal government govern the support Army forces provide in response to disaster or a damaging attack on the homeland. In the event of disaster or attack, Army forces support civil authorities with essential services including:

- Performing rescues.
- Providing emergency medical care.
- Preventing epidemic disease.
- Providing food and water.
- Providing emergency shelter.
- Providing basic sanitation (sewage and garbage disposal).
- Providing minimum essential access to affected areas.

Army forces work directly with state and Federal officials to help restore and return control of services to civil authorities as rapidly as possible. Any disaster or attack will reduce the state and local governments' capacities. Army forces provide support as command and control, protection, and sustainment to government agencies at all levels until they can carry out their responsibilities without Army assistance.

3-102. Although the risk of an attack by a foreign power is low, and likely to remain so given the current preponderance of U.S. military power, this is not the case when planning for a possible major terror attack. As the terrible events of 11 September 2001 demonstrated, even the most powerful military deterrent is not enough to prevent stateless terrorism. In the worst case scenario, an attacker might use a weapon of mass destruction. In the event of an attack on the United States or its territories, the first responsibility of Army forces will be to provide civil authorities with capabilities to save lives and protect critical infrastructure from further attacks. If civilian law enforcement agencies cannot maintain law and order, Army forces may be directed to assist in that capacity.

Support Civil Law Enforcement

3-103. When authorized and directed, Army forces provide support to local, state, and Federal law enforcement officers. In extreme cases, and when directed by the President, Regular Army forces maintain law and order.

Provide Other Support as Required

3-104. The Army is frequently called upon to provide other support to civil authorities apart from disaster response and law enforcement. Most of this support is identified well in advance and planned with civil authorities. Much of the support is routine; it consists of providing support to communities surrounding Army units' home stations. Examples of other types of support provided to civil authorities include the following:

- Support to state funerals.
- Participation in major public sporting events.
- Providing military equipment and Soldiers to community events.

Purposes of Civil Support Tasks

3-105. Army forces execute civil support tasks for the following purposes.

Save Lives

3-106. The first priority in civil support operations is to save lives. In the aftermath of a man-made or natural disaster, the first military forces to arrive focus on rescue, evacuation, and consequence management.

Restore Essential Services

3-107. In any major disaster, citizens suffer and may die because most, if not all, essential services are disrupted. This disruption leads to tremendous suffering and the spread of disease. Restoring essential services is crucial to saving lives over the long term and providing the first step to recovery.

Maintain or Restore Law and Order

3-108. When authorized, Army forces assist local, state, and Federal authorities with law enforcement. Often the support is provided under crisis conditions when events overwhelm civil capacity. In other cases, the Army provides personnel and equipment to support ongoing law enforcement activities, such as control of U.S. borders. In all instances, Army forces use lethal force in accordance with rules for the use of force and only as a last resort.

Protect Infrastructure and Property

3-109. In the aftermath of a disaster or civil disturbance, Army forces frequently secure public and private property. This allows civilian law enforcement to focus on dealing with criminal behavior.

Maintain or Restore Local Government

3-110. In a disaster, local government may be unable to carry out its normal functions. Army forces provide essential services and communications support to government officials until they can resume their normal functions.

Shape the Environment for Interagency Success

3-111. Success in civil support operations is measured by the success of civilian officials in carrying out their responsibilities. Civil support helps government officials meet their responsibilities to the public, ultimately without assistance from military forces.

Civil Support and Homeland Security

3-112. Army forces conduct civil support operations as part of homeland security. Homeland security provides the Nation with strategic flexibility by protecting its citizens, critical assets, and infrastructure from conventional and unconventional threats. It includes the following missions. (JP 3-28 discusses these missions in detail.)

Homeland Defense

3-113. Homeland defense protects the United States from direct attack or a threat by hostile armed forces. In the event of such an attack, Army forces under joint command conduct offensive and defensive operations against the enemy while providing civil support to Federal authorities. A defensive task routinely conducted in homeland defense missions is protecting critical assets and key infrastructure during crises. The ability to conduct offensive operations, though maintained primarily as a potential, is also present.

Civil Support

3-114. Civil support includes the key tasks of providing support in response to disaster and supporting law enforcement (as discussed above). Unless the Nation is attacked, Army forces conduct civil support operations exclusive of the offense and defense.

Emergency Preparedness Planning

3-115. In emergency preparedness planning, Department of Homeland Security examines a wide range of threats and plans for man-made and natural disasters and incidents. Department of Defense supports emergency preparedness planning. When necessary, these plans are executed as civil support operations.

COMBINING THE ELEMENTS OF FULL SPECTRUM OPERATIONS

3-116. Within the concept of operations, the proportion and role of offensive, defensive, and stability or civil support tasks varies based on several factors. Changes in the nature of the operation, the tactics used, and where the environment falls on the spectrum of conflict affect the mix and focus. Some combinations may be sequential, such as a mobile defense followed by a counteroffensive, but many occur simultaneously. During major combat operations, a division may be attacking in one area, defending in another, and focusing on stability tasks in a third. Offensive and defensive operations may be complemented with stability tasks and vice versa at any point of a campaign. Simultaneous combinations are also present in operational themes dominated by stability. A peace operation, for example, may include a mix of several elements. One force may be conducting a raid against hostile forces (offense), while a second is securing an important airport (defense), and a third is providing sanitary and secure facilities to dislocated civilians (stability). In homeland security, civil support is often the only element executed, although there may be some planning for defense.

3-117. Differing combinations of the elements of full spectrum operations generally characterize each operational theme. (See figure 3-4, page 3-20.) The combinations vary according to the conditions and requirements for each phase of a campaign or major operation. Commanders determine the weight of effort by considering the primary tasks and purposes for each element within the operational theme and which will be decisive. This allows commanders to translate their design into tactical actions.

3-118. Conducting full spectrum operations involves more than simultaneous execution of all its elements. It requires commanders and staffs to consider their units' capabilities and capacities relative to each element. Commanders consider their missions, decide which tactics to use, and balance the elements of full spectrum operations while preparing their concept of operations. They determine which tasks the force can accomplish simultaneously, if phasing is required, what additional resources may be necessary, and how to transition from one element to another. At the operational level, this requires looking beyond the current operation and prioritizing each element for the next phase or sequel.

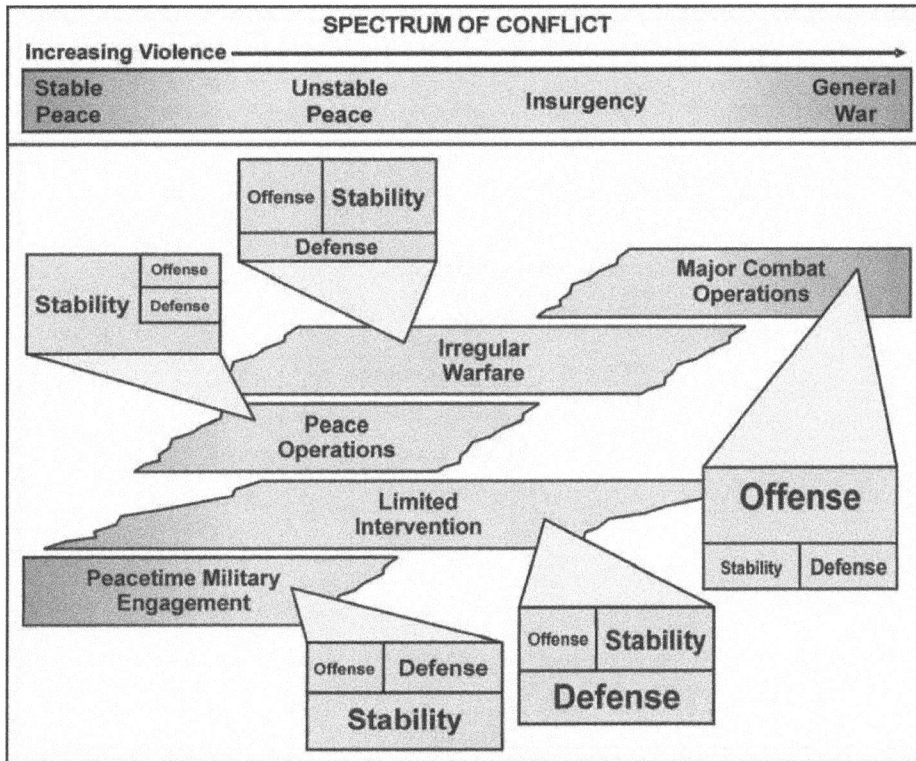

Figure 3-4. Examples of combining the elements of full spectrum operations within operational themes

3-119.   The transition between elements of full spectrum operations requires careful assessment, prior planning, and unit preparation as commanders shift their combinations of full spectrum operations. This begins with an assessment of the situation to determine which primary tasks are applicable and the priority for each. For example, a division assigns a brigade combat team an area of operations and the tasks of eliminating any enemy remnants, securing a dam, and conducting stability operations following a joint offensive phase. The brigade commander determines that the brigade will conduct three tasks: an area defense of the dam, control the civil population in the area and exclude all civilians from the area of the dam, and conduct movements to contact in various objective areas, specifically search and attack operations. Simultaneously, the brigade staff begins planning for the next phase in which civil security, civil control, and assisting the local authorities with essential services will become priorities, while continuing to defend the dam. Reconnaissance and surveillance, joint information operations, area and route security operations, and protection are continuous. The commander assigns tasks to subordinates, modifies the brigade task organization, replenishes, and requests additional resources if required. Depending on the length of operations, the higher headquarters may establish unit training programs to prepare units for certain tasks.

3-120.   When conditions change, commanders adjust the combination of the elements of full spectrum operations in the concept of operations. When an operation is phased, these changes are included in the plan. The relative weight given to each element varies with the actual or anticipated conditions. It is reflected in tasks assigned to subordinates, resource allocation, and task organization. Full spectrum operations is not a phasing method. Commanders consider the concurrent conduct of each element—offense, defense, and stability or civil support—in every phase of an operation. Figure 3-5 (based on a similar figure in JP 3-0), illustrates combinations and weighting of the elements of full spectrum operations across the phases of a campaign. The phases are examples. An actual campaign may name and array phases differently. (JP 3-0 discusses the campaign phases.)

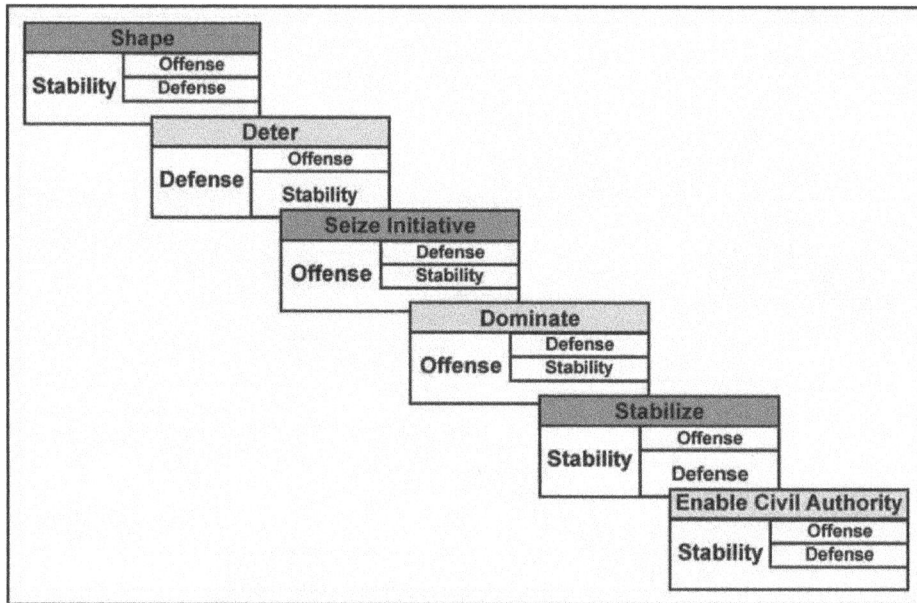

Figure 3-5. Examples of combining the elements of full spectrum operations in a notional campaign

3-121. Seizing, retaining, and exploiting the initiative requires commanders to interpret developments and shift the weight of effort throughout their operations to achieve decisive results. As they do this, the forces and priorities they assign to each element of full spectrum operations change. Throughout an operation, commanders constantly adapt and perform many tasks simultaneously. Commanders change tactics, modify their command and control methods, change task organization, and adjust the weight placed on each element of full spectrum operations to keep the force focused on accomplishing the mission. They base these decisions on the situation, available resources, and the force's ability to execute multiple, diverse tasks. Commanders' assessments should consider the progress of ongoing operations, changes in the situation, and how the rules of engagement affect the force's effectiveness in each element. Commanders not only assess how well a current operation is accomplishing the mission but also how its conduct is shaping the situation for subsequent missions.

3-122. Applying tactical and operational art in full spectrum operations involves knowing when and if simultaneous combinations are appropriate and feasible. Every operation does not require offensive tasks; stability or civil support may be the only elements executed. Nonetheless, commanders and staffs always consider each element of full spectrum operations and its relevance to the situation. An element may be unnecessary, but it is the commander who determines that. Not every echelon or unit necessarily executes simultaneous full spectrum operations. Division and higher echelon operations normally combine three elements simultaneously. Brigade combat teams may focus exclusively on a single element when attacking or defending, shifting priority to another element as the plan or situation requires. Battalion and smaller units often execute the elements sequentially, based on their capabilities and the situation. However, simultaneous execution of offensive, defensive, and stability tasks at lower echelons is common in irregular warfare and peace operations. Sometimes the force available is inadequate for executing simultaneous combinations of offensive, defensive, and stability tasks. In these cases, commanders inform the higher headquarters of the requirement for additional forces. Regardless of the situation, commanders assess the risk to units and mission accomplishment. Combining elements of full spectrum operations requires the following:

- A clear concept of operations that establishes the role of each element and how it contributes to accomplishing the mission.
- A flexible command and control system.
- Clear situational understanding.
- Aggressive intelligence gathering and analysis.
- Aggressive security operations.
- Units that can quickly change their task organization.
- An ability to respond quickly.
- Responsive sustainment.
- Combat power applied through combined arms, including applicable joint capabilities. (See chapter 4.)

SUMMARY

3-123. The Army's operational concept, full spectrum operations, describes how Army forces conduct operations. The complex nature of the operational environment requires commanders to simultaneously combine offensive, defensive, and stability or civil support tasks to accomplish missions domestically and abroad. Each element of full spectrum operations includes a basic set of tasks and related purposes. Mission command directs the application of full spectrum operations to seize, retain, and exploit the initiative and achieve decisive results.

Chapter 4

Combat Power

This chapter discusses the nature of combat power and how Army forces use the warfighting functions to generate combat power. The eight elements of combat power include the six warfighting functions—movement and maneuver, intelligence, fires, sustainment, command and control, and protection—multiplied by leadership and complemented by information. Commanders use combined arms to increase the effects of combat power through complementary and reinforcing capabilities. Army forces achieve combined arms through force tailoring, task organization, and mutual support.

THE ELEMENTS OF COMBAT POWER

4-1.   Full spectrum operations require continuously generating and applying combat power, often for extended periods. *Combat power* **is the total means of destructive, constructive, and information capabilities that a military unit/formation can apply at a given time. Army forces generate combat power by converting potential into effective action.**

4-2.   Commanders conceptualize capabilities in terms of combat power. There are eight elements of combat power. These are leadership, information, movement and maneuver, intelligence, fires, sustainment, command and control, and protection. Leadership and information are applied through, and multiply the effects of, the other six elements of combat power. These six—movement and maneuver, intelligence, fires, sustainment, command and control, and protection—are collectively described as the warfighting functions. Commanders apply combat power through the warfighting functions using leadership and information. (See figure 4-1.)

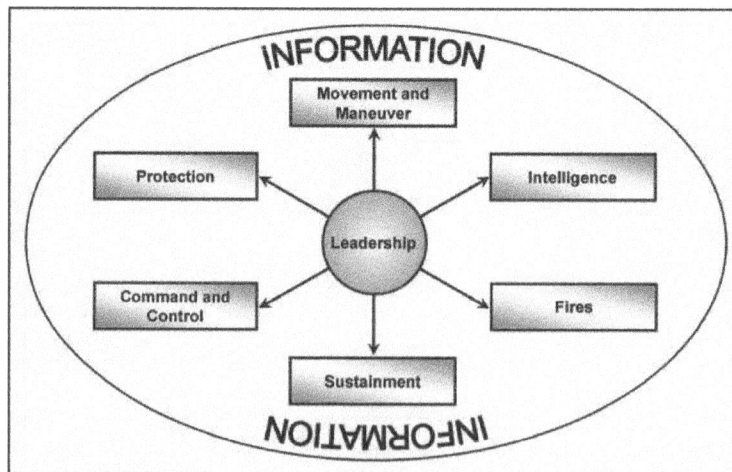

Figure 4-1. The elements of combat power

4-3. In full spectrum operations, every unit—regardless of type—either generates or maintains combat power. All contribute to operations. Commanders ensure deployed Army forces have enough potential combat power to combine the elements of full spectrum operations in ways appropriate to conditions. Ultimately, Army forces combine elements of combat power to defeat the enemy and master situations.

4-4. Commanders balance the ability to mass lethal and nonlethal effects with the need to deploy and sustain the units that produce those effects. They balance accomplishing the mission quickly with being able to project and sustain the force. Generating and maintaining combat power throughout an operation is essential to success. Commanders tailor force packages to maximize the capability of the initial-entry force. Follow-on units increase endurance and ability to operate in depth. Many factors contribute to generating combat power. These include the following:

- Employing reserves.
- Focusing joint support.
- Rotating committed forces.
- Staging sustainment assets to preserve momentum and synchronization.

4-5. Combat power is not a numerical value. It can be estimated but not quantified. Combat power is always relative. It has meaning only in relation to conditions and enemy capabilities. It is relevant solely at the point in time and space where it is applied. In addition, how an enemy generates and applies combat power may fundamentally differ from that of Army forces. It is dangerous to assume that enemy capabilities are a mirror image of friendly capabilities. Before an operation, combat power is unrealized potential. Through leadership, this potential is transformed into action. Commanders use information to integrate and enhance action. Information is also applied through the warfighting functions to shape the operational environment and complement action. Combat power becomes decisive when applied by skilled commanders leading well-trained Soldiers and units. Ultimately, commanders achieve success by applying superior combat power at the decisive place and time.

LEADERSHIP

4-6. Leadership is applied through Soldiers to the warfighting functions. Leadership is the multiplying and unifying element of combat power. Confident, competent, and informed leadership intensifies the effectiveness of all other elements of combat power by formulating sound operational ideas and assuring discipline and motivation in the force. Good leaders are the catalyst for success. Effective leadership can compensate for deficiencies in all the warfighting functions because it is the most dynamic element of combat power. The opposite is also true; poor leadership can negate advantages in warfighting capabilities. The Army defines *leadership* as the process of influencing people by providing purpose, direction, and motivation, while operating to accomplish the mission and improving the organization (FM 6-22). An Army leader, by virtue of assumed role or assigned responsibility, inspires and influences people to accomplish organizational goals. Army leaders motivate people to pursue actions, focus thinking, and shape decisions for the greater good of the organization. They instill in Soldiers the will to win. Army doctrine describes essential leadership attributes (character, presence, and intellect) and competencies (lead, develop, and achieve). These attributes and competencies mature through lifelong learning. (FM 6-22 contains Army leadership doctrine.)

4-7. Leaders influence not only Soldiers but other people as well. Leadership is crucial in dealing with civilians in any conflict or disaster. Face-to-face contact with people in the area of operations encourages cooperation between civilians and Soldiers. Army leaders work with members of other Services and civilian organizations. These leaders strive for the willing cooperation of multinational military and civilian partners. The Army requires self-aware, adaptive leaders who can both defeat the enemy in combat and master complexities of operations dominated by stability or civil support.

4-8. Leadership in today's operational environment is often the difference between success and failure. Leaders provide purpose, direction, and motivation in all operations. Through training and by example, leaders develop cultural awareness in Soldiers. This characteristic improves Soldiers' ability to cope with the ambiguities of complex environments. Leadership ensures Soldiers understand the purpose of operations and use their full capabilities. In every operation, Army leaders clarify purpose and mission, direct operations, and set the example for courage and competence. They hold their Soldiers to the Army Values and ensure their Soldiers comply with the law of war.

INFORMATION

4-9. Information is a powerful tool in the operational environment. In modern conflict, information has become as important as lethal action in determining the outcome of operations. Every engagement, battle, and major operation requires complementary information operations to both inform a global audience and to influence audiences within the operational area; it is a weapon against enemy command and control and is a means to affect enemy morale. It is both destructive and constructive. Commanders use information to understand, visualize, describe, and direct the warfighting functions. Soldiers constantly use information to persuade and inform target audiences. They also depend on data and information to increase the effectiveness of the warfighting functions.

4-10. Since information shapes the perceptions of the civilian population, it also shapes much of the operational environment. All parties in a conflict use information to convey their message to various audiences. These include enemy forces, adversaries, and neutral and friendly populations. Information is critical in stability operations where the population is a major factor in success. While the five stability tasks are essential for success, without complementary information engagement that explains these actions to the population, success may be unattainable. Information must be proactive as well as reactive. The enemy adeptly manipulates information and combines message and action effectively. Countering enemy messages with factual and effective friendly messages can be as important as the physical actions of Soldiers. The effects of each warfighting function should complement information objectives (the message) while information objectives stay consistent with Soldiers' actions.

4-11. The joint force continues to modernize information systems. These improvements provide leaders with the information necessary to enhance and focus the warfighting functions. Leadership based on relevant information enables the commander to make informed decisions on how best to apply combat power. Ultimately, this creates opportunities to achieve decisive results. The computer-displayed common operational picture is a good example. It provides commanders with improved situational awareness by merging a lot of information into displays that Soldiers can understand at a glance. Information disseminated by information systems allows leaders at all levels to make better decisions quickly. The common operational picture lets Army forces use lethal and nonlethal actions more effectively than the enemy can. For example, accurate intelligence disseminated quickly by information systems allows friendly forces to maneuver around enemy engagement areas while massing the effects of combat power at the decisive place and time. This reduces friendly casualties and may allow a small force to defeat a larger enemy force.

4-12. Commanders use the concept of information superiority (see chapter 7) and its four information-related contributors to understand and visualize the role of information in their operational environment. They use various information capabilities to shape the operational environment.

WARFIGHTING FUNCTIONS

4-13. Commanders use the warfighting functions to help them exercise battle command. **A *warfighting function* is a group of tasks and systems (people, organizations, information, and processes) united by a common purpose that commanders use to accomplish missions and training objectives**. Decisive, shaping, and sustaining operations combine all the warfighting functions to generate combat power. (See paragraphs 5-59 through 5-64.) No warfighting function is exclusively decisive, shaping, or sustaining. The Army's warfighting functions are fundamentally linked to the joint functions. They also parallel those of the Marine Corps.

MOVEMENT AND MANEUVER

4-14. The *movement and maneuver warfighting function* is the related tasks and systems that move forces to achieve a position of advantage in relation to the enemy. Direct fire is inherent in maneuver, as is close combat. The function includes tasks associated with force projection related to gaining a positional advantage over an enemy. One example is moving forces to execute a large-scale air or airborne assault. Another is deploying forces to intermediate staging bases in preparation for an offensive. *Maneuver* is the employment of forces in the operational area through movement in combination with fires to achieve a position of advantage in respect to the enemy in order to accomplish the mission (JP 3-0). Maneuver is the means by which commanders mass the effects of combat power to achieve surprise, shock, and momentum. Effective maneuver requires close coordination with fires. Movement is necessary to disperse and displace the force as a whole or in part when maneuvering. Both tactical and operational maneuver require logistic support. The movement and maneuver warfighting function includes the following tasks:

- Deploy.
- Move.
- Maneuver.
- Employ direct fires.
- Occupy an area.
- Conduct mobility and countermobility operations.
- Employ battlefield obscuration.

The movement and maneuver warfighting function does not include administrative movements of personnel and materiel. These movements fall under the sustainment warfighting function.

4-15. FM 3-90 discusses maneuver and tactical movement. FMI 3-35 and FMs 100-17-1 and 100-17-2 discuss force projection.

INTELLIGENCE

4-16. The *intelligence warfighting function* **is the related tasks and systems that facilitate understanding of the operational environment, enemy, terrain, and civil considerations**. It includes tasks associated with intelligence, surveillance, and reconnaissance (ISR) operations, and is driven by the commander. (See chapter 7.) Intelligence is more than just collection. It is a continuous process that involves analyzing information from all sources and conducting operations to develop the situation. The intelligence warfighting function includes the following tasks:

- Support to force generation.
- Support to situational understanding.
- Conduct ISR.
- Provide intelligence support to targeting and information capabilities.

4-17. FM 2-0 describes the intelligence warfighting function. Several unit-level manuals provide supplemental doctrine on surveillance and reconnaissance.

FIRES

4-18. The *fires warfighting function* **is the related tasks and systems that provide collective and coordinated use of Army indirect fires, joint fires, and command and control warfare, including nonlethal fires, through the targeting process**. It includes tasks associated with integrating and synchronizing the effects of these types of fires and command and control warfare—including nonlethal fires—with the effects of other warfighting functions. These are integrated into the concept of operations during planning and adjusted based on the targeting guidance. Fires normally contribute to the overall effect of maneuver but commanders may use them separately for the decisive operation and shaping operations. The fires warfighting function includes the following tasks:

- Decide surface targets.
- Detect and locate surface targets.
- Provide fire support.
- Assess effectiveness.
- Integrate command and control warfare, including nonlethal fires.

4-19. Paragraphs 7-23 through 7-30 discuss command and control warfare. When revised, FM 3-13 will address command and control warfare.

SUSTAINMENT

4-20. **The *sustainment warfighting function* is the related tasks and systems that provide support and services to ensure freedom of action, extend operational reach, and prolong endurance**. The endurance of Army forces is primarily a function of their sustainment. Sustainment determines the depth and duration of Army operations. It is essential to retaining and exploiting the initiative. Sustainment is the provision of the logistics, personnel services, and health service support necessary to maintain operations until mission accomplishment. Internment, resettlement, and detainee operations fall under the sustainment warfighting function and include elements of all three major subfunctions. FM 4-0 describes the sustainment warfighting function.

Logistics

4-21. *Logistics* is the science of planning and carrying out the movement and maintenance of forces. In its most comprehensive sense, those aspects of military operations that deal with: a. design and development, acquisition, storage, movement, distribution, maintenance, evacuation, and disposition of materiel; b. movement, evacuation, and hospitalization of personnel; c. acquisition or construction, maintenance, operation, and disposition of facilities; and d. acquisition or furnishing of services (JP 1-02). Although joint doctrine defines it as science, logistics involves both military art and science. Knowing when and how to accept risk, prioritizing myriad requirements, and balancing limited resources all require military art. Logistics integrates strategic, operational, and tactical support of deployed forces while scheduling the mobilization and deployment of additional forces and materiel. Logistics includes—

- Maintenance.
- Transportation.
- Supply.
- Field services.
- Distribution.
- Contracting.
- General engineering support.

Personnel Services

4-22. Personnel services are those sustainment functions related to Soldiers' welfare, readiness, and quality of life. Personnel services complement logistics by planning for and coordinating efforts that provide and sustain personnel. Personnel services include—

- Human resources support.
- Financial management.
- Legal support.
- Religious support.
- Band support.

Health Service Support

4-23. Health service support consists of all support and services performed, provided, and arranged by the Army Medical Department. It promotes, improves, conserves, or restores the mental and physical well-being of Soldiers and, as directed, other personnel. This includes casualty care, which involves all Army Medical Department functions, including—

- Organic and area medical support.
- Hospitalization.
- Dental care.

- Behavioral health and neuropsychiatric treatment.
- Clinical laboratory services and treatment of chemical, biological, radiological, and nuclear patients.
- Medical evacuation.
- Medical logistics.

Health service support closely relates to force health protection. (See paragraph 4-28.)

COMMAND AND CONTROL

4-24. The *command and control warfighting function* **is the related tasks and systems that support commanders in exercising authority and direction**. It includes those tasks associated with acquiring friendly information, managing relevant information, and directing and leading subordinates. Through command and control, commanders integrate all warfighting functions to accomplish the mission. The command and control warfighting function includes the following tasks:

- Execute the operations process (see chapter 5).
- Conduct command post operations.
- Integrate the information superiority contributors—the Army information tasks, ISR, knowledge management, and information management.
- Conduct information engagement.
- Conduct civil affairs activities.
- Integrate airspace command and control.
- Execute command programs.

4-25. The command and control warfighting function is the primary integrator of information tasks associated with information superiority. Army leaders use information tasks to shape the operational environment throughout the operations process. During operations against an enemy, they degrade the enemy's ability to do the same.

4-26. Chapter 5 provides additional discussion of command and control and the operations process. Chapter 7 discusses information superiority. FM 6-0 describes command and control warfighting doctrine.

PROTECTION

4-27. The *protection warfighting function* **is the related tasks and systems that preserve the force so the commander can apply maximum combat power**. Preserving the force includes protecting personnel (combatants and noncombatants), physical assets, and information of the United States and multinational military and civilian partners. The protection warfighting function facilitates the commander's ability to maintain the force's integrity and combat power. Protection determines the degree to which potential threats can disrupt operations and counters or mitigates those threats. Emphasis on protection increases during preparation and continues throughout execution. Protection is a continuing activity; it integrates all protection capabilities to safeguard bases, secure routes, and protect forces. The protection warfighting function includes the following tasks:

- Air and missile defense.
- Personnel recovery.
- Information protection.
- Fratricide avoidance.
- Operational area security.
- Antiterrorism.
- Survivability.
- Force health protection.
- Chemical, biological, radiological, and nuclear operations.
- Safety.

- Operations security.
- Explosive ordnance disposal.

4-28. The protection warfighting function includes force health protection. Force health protection includes all measures to promote, improve, or conserve the mental and physical well-being of Soldiers. These measures enable a healthy and fit force, prevent injury and illness, and protect the force from health hazards. The measures include the prevention aspects of a number of Army Medical Department functions, including the following:

- Preventive medicine, including—
  - Medical surveillance.
  - Occupational and environmental health surveillance.
- Veterinary services, including—
  - Food inspection.
  - Animal care.
  - Prevention of zoonotic diseases (those transmitted from animals to humans, such as plague or rabies).
- Combat and operational stress control.
- Dental services (preventive dentistry).
- Laboratory services (area medical laboratory support).

4-29. When revised, FM 3-13 will address information protection.

COMBINED ARMS

4-30. Applying combat power depends on combined arms to achieve its full destructive, disruptive, informational, and constructive potential. **Combined arms is the synchronized and simultaneous application of the elements of combat power to achieve an effect greater than if each element of combat power was used separately or sequentially.** Combined arms merges leadership, information, and each of the warfighting functions and their supporting systems. Used destructively, combined arms integrates different capabilities so that counteracting one makes the enemy vulnerable to another. Used constructively, combined arms multiplies the effectiveness and the efficiency of Army capabilities used in stability or civil support.

4-31. Combined arms uses the capabilities of each warfighting function and information in complementary and reinforcing capabilities. *Complementary* capabilities protect the weaknesses of one system or organization with the capabilities of a different warfighting function. For example, commanders use artillery (fires) to suppress an enemy bunker complex pinning down an infantry unit (movement and maneuver). The infantry unit then closes with and destroys the enemy. In this example, the fires warfighting function complements the maneuver warfighting function. *Reinforcing* capabilities combine similar systems or capabilities within the same warfighting function to increase the function's overall capabilities. In urban operations, for example, infantry, aviation, and armor (movement and maneuver) often operate closely together. This combination reinforces the protection, maneuver, and direct fire capabilities of each. The infantry protects tanks from enemy infantry and antitank systems; tanks provide protection and firepower for the infantry. Attack helicopters maneuver freely above buildings to fire from positions of advantage, while other aircraft help sustain the ground elements. Together, these capabilities form a lethal team built on movement and maneuver. In another example, multiple artillery units routinely mass fires in support of a committed artillery battalion (reinforcement). Joint capabilities, such as close air support and special operations forces, can complement or reinforce Army forces' capabilities.

4-32. Combined arms operations are familiar to Army forces. Unified actions—those integrating the capabilities of joint forces with those of multinational military and civilian organizations—have become typical as well. This integration requires careful preparation. Training and exchange of liaison at every level are necessary for successful unified action.

4-33. Combined arms multiplies Army forces' effectiveness in all operations. Units operating without support of other capabilities generate less combat power and may not accomplish their mission. Employing combined arms requires highly trained Soldiers, skilled leadership, effective staff work, and integrated information systems. Commanders synchronize combined arms to apply the effects of combat power to best advantage. The sequence and simultaneity of combined arms actions vary with both the operational or tactical design and in execution. Typically, intelligence, surveillance, and reconnaissance activities begin soon after receipt of mission and continue throughout preparation and execution. They do not cease after mission accomplishment but continue as needed. Sustainment and protection are conducted constantly but may peak before and after execution. Maneuver and fires complement each other continuously but sometimes precede each other. For example, the commander conducts preparatory lethal fires combined with nonlethal electronic warfare to isolate and destroy enemy forces on an objective before maneuver forces make contact. Another example is the shifting of fires beyond the immediate vicinity of maneuver units during a pursuit.

4-34. Combined arms is achieved through organizational design and temporary reorganization (tailored and task-organized forces). For example, units organic to brigade combat teams perform all warfighting functions. However, the capabilities organic to the brigade combat team do not include Army aviation, air and missile defense, and Army special operations forces. When required, these capabilities are added through force tailoring and task organization. Higher echelons achieve combined arms capabilities by tailoring and task-organizing different types of brigades and battalions under corps or division headquarters. For example, a division force commander may complement or reinforce four or five brigade combat teams with any or all of the modular support brigades and functional brigades described in appendix C.

4-35. Irregular warfare, including counterinsurgency, requires a combined arms approach emphasizing small-unit capabilities and information tasks. Irregular warfare also requires a mix of forces different from conventional warfare to address the fundamental differences between them. Since irregular warfare concerns the struggle for control and influence over a population, task organizations need to reflect this. Furthermore, requirements that characterize counterinsurgency in general, and civil security and civil control in particular, vary significantly among tactical-level areas of operations. This situation requires releasing intelligence, civil affairs, and information assets typically held at higher headquarters to brigade combat teams and often to battalion task forces. Unified action at lower echelons characterizes irregular warfare. Liaison officers and adjacent unit coordination are essential to integrating and synchronizing Army operations with those of other organizations. Higher headquarters may need to reinforce the command and control capabilities of their subordinates to improve coordination between them and the various organizations in their areas of operations.

FORCE TAILORING

4-36. **Force tailoring is the process of determining the right mix of forces and the sequence of their deployment in support of a joint force commander**. It involves selecting the right force structure for a joint operation from available units within a combatant command or from the Army force pool. The selected forces are then sequenced into the operational area as part of force projection. Joint force commanders request and receive forces for each campaign phase, adjusting the quantity and Service component of forces to match the weight of effort required. Army Service component commanders tailor Army forces to meet land force requirements determined by joint force commanders. Army Service component commanders also recommend forces and a deployment sequence to meet those requirements. Force tailoring is continuous. As new forces rotate into the operational area, forces with excess capabilities return to the supporting combatant and Army Service component commands. (Paragraphs B-3 through B-4 and C-8 through C-10 discuss Army Service component commands.)

TASK-ORGANIZING

4-37. **Task-organizing is act of designing an operating force, support staff, or logistic package of specific size and composition to meet a unique task or mission. Characteristics to examine when task-organizing the force include, but are not limited to: training, experience, equipage, sustainability, operating environment, enemy threat, and mobility. For Army forces, it includes allocating available** assets to subordinate commanders and establishing their command and support relationships. Task-organizing occurs within a tailored force package as commanders organize groups of units for specific missions. It continues as commanders reorganize units for subsequent missions. The ability of Army forces to task-organize gives them extraordinary agility. It lets operational and tactical commanders configure their units to best use available resources. It also allows Army forces to match unit capabilities rapidly to the priority assigned to offensive, defensive, and stability or civil support tasks.

MUTUAL SUPPORT

4-38. Commanders consider mutual support when task-organizing forces and assigning areas of operations and positioning units. *Mutual support* is that support which units render each other against an enemy, because of their assigned tasks, their position relative to each other and to the enemy, and their inherent capabilities (JP 1-02). In Army doctrine, mutual support is a planning consideration related to force disposition, not a command relationship. (See appendix B.) Mutual support has two aspects—supporting range and supporting distance. (See figure 4-2.) Understanding mutual support and accepting risk during operations are fundamental to the art of tactics.

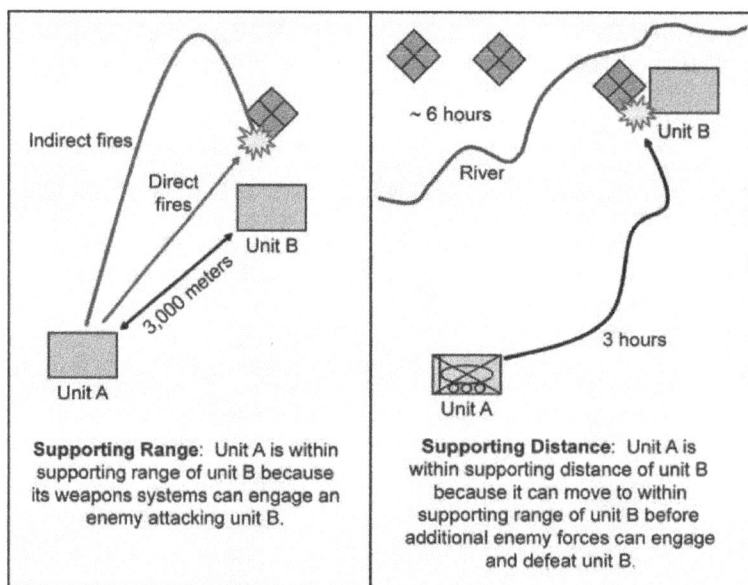

Figure 4-2. Examples of supporting range and supporting distance

4-39. *Supporting range* **is the distance one unit may be geographically separated from a second unit yet remain within the maximum range of the second unit's weapons systems.** It depends on available weapons systems and is normally the maximum range of the supporting unit's indirect fire weapons. For small units (such as squads, sections, and platoons), it is the distance between two units that their direct fires can cover effectively. Supporting range may be limited by visibility; if one unit cannot effectively or safely fire in support of another, the first may not be in supporting range even though its weapons have the required range.

4-40. *Supporting distance* **is the distance between two units that can be traveled in time for one to come to the aid of the other and prevent its defeat by an enemy or ensure it regains control of a civil situation.** The following factors affect supporting distance:

- Terrain and mobility.
- Distance.
- Enemy capabilities.

- Friendly capabilities.
- Reaction time.

When friendly forces are static, supporting range equals supporting distance.

4-41. Supporting distance and supporting range are affected by the command and control capabilities of supported and supporting units. Units may be within supporting distance, but if the supported unit cannot communicate with the supporting unit, the supporting unit may not be able to affect the operation's outcome. In such cases, the units are not within supporting distance, regardless of their proximity to each other. If the units share a common operational picture, the situation may be quite different. Relative proximity may be less important than both units' ability to coordinate their maneuver and fires. To exploit the advantage of supporting distance, the units must be able to synchronize their maneuver and fires more effectively than the enemy can. Otherwise, the enemy may be able to defeat both units in detail.

4-42. Commanders also consider supporting distance in operations dominated by stability or civil support. Units maintain mutual support when one unit can draw on another's capabilities. An interdependent joint force may make proximity less significant than available capability. For example, Air Force assets may be able to move a preventive medicine detachment from an intermediate staging base to an operational area threatened by an epidemic. Additional treatment capability might be moved to the operational area based on the threat to Soldiers and the populace.

4-43. Improved access to joint capabilities gives commanders additional means to achieve mutual support. Those capabilities can extend the operating distances between Army units. Army commanders can substitute joint capabilities for mutual support between subordinate forces. Doing this multiplies supporting distance many times over. Army forces can then extend operations over greater areas at a higher tempo. Joint capabilities are especially useful when subordinate units operate in noncontiguous areas of operations that place units beyond supporting distance or supporting range. However, depending on them entails accepting risk.

SUMMARY

4-44. The elements of combat power consist of six warfighting functions tied together by leadership and enhanced by information. Army forces apply combat power most effectively through combined arms. Combined arms generates more combat power than employing arms individually. Army forces are organized for combined arms using force tailoring and task organization. Mutual support used in reinforcing and complementary combinations also multiples the effects of combat power.

Chapter 5

Command and Control

This chapter describes how commanders exercise command and control throughout the operations process (planning, preparation, execution, and continuous assessment). It presents battle command in terms of understanding, visualizing, describing, directing, leading, and assessing. The chapter discusses control measures and the common operational picture. It concludes with a description of the operations process, including integrating processes and continuing activities. Mission command permeates the exercise of command and control. Commanders use mission command to create a positive command climate that fosters trust and mutual understanding and encourages opportunistic actions by subordinates.

EXERCISE OF COMMAND AND CONTROL

5-1. *Command and control* is the exercise of authority and direction by a properly designated commander over assigned and attached forces in the accomplishment of a mission. Commanders perform command and control functions through a command and control system (FM 6-0). Command and control is fundamental to the art and science of warfare. Each warfighting function relies on it for integration and synchronization. Command and control uses both art and science. Commanders combine the art of command and the science of control to accomplish missions.

5-2. The commander is the focus of command and control. Through it commanders assess the situation, make decisions, and direct actions. However, commanders cannot exercise command and control alone except in the smallest organizations. Thus, commanders perform these functions through a *command and control system*—the arrangement of personnel, information management, procedures, and equipment and facilities essential for the commander to conduct operations (FM 6-0). An effective command and control system is essential for commanders to conduct (plan, prepare, execute, and assess) operations that accomplish the mission decisively.

5-3. Trained and disciplined Soldiers are the single most important element of any command and control system. Their actions and responses—everything from fire and maneuver techniques to the disciplined observation of rules of engagement—are means of controlling operations. Soldiers also assist commanders and exercise control on their behalf. Staffs perform many functions that help commanders exercise command and control. These include—

- Providing relevant information and analysis.
- Maintaining running estimates (see paragraph 5-89) and making recommendations.
- Preparing plans and orders.
- Monitoring operations.
- Controlling operations.
- Assessing the progress of operations.

COMMAND

5-4. Command and control are interrelated. *Command* is the authority that a commander in the armed forces lawfully exercises over subordinates by virtue of rank or assignment. Command includes the authority and responsibility for effectively using available resources and for planning the employment of, organizing, directing, coordinating, and controlling military forces for the accomplishment of assigned missions. It also includes responsibility for health, welfare, morale, and discipline of assigned personnel (JP 1). Leaders possessing command authority strive to use it with firmness, care, and skill. The authority of command provides the basis for control.

5-5.   Command is an individual and personal function. It blends imaginative problem solving, motivational and communications skills, and a thorough understanding of the dynamics of operations. Command during operations requires understanding the complex, dynamic relationships among friendly forces, enemies, and the environment, including the populace. This understanding helps commanders visualize and describe their commander's intent and develop focused planning guidance. (Paragraph 5-55 defines commander's intent.)

CONTROL

5-6.   While command is a personal function, control involves the entire force. **Control is the regulation of forces and warfighting functions to accomplish the mission in accordance with the commander's intent**. It is fundamental to directing operations. Commanders and staffs both exercise control. Aided by staffs, commanders exercise control over all forces in their area of operations. Staffs coordinate actions, keep the commander informed, and exercise control for the commander.

5-7.   Commanders and staffs must understand the science of control to overcome the physical and procedural constraints under which units operate. Control demands understanding those aspects of operations that can be analyzed and measured. It relies on objectivity, facts, empirical methods, and analysis. These include the physical capabilities and limitations of friendly and enemy organizations and systems. Control also requires a realistic appreciation for time-distance factors and the time required to initiate certain actions. The science of control includes the tactics, techniques, and procedures used to accomplish command and control tasks. It includes operational terms and graphics.

BATTLE COMMAND

5-8.   In battle, commanders face a thinking and adaptive enemy. Commanders estimate, but cannot predict, the enemy's actions and the course of future events. Two key concepts for exercising command and control in operations are battle command and mission command. Battle command describes the commander's role in the operations process. Mission command is the Army's preferred means of battle command. (See paragraphs 3-29 through 3-35.)

5-9.   **Battle command is the art and science of understanding, visualizing, describing, directing, leading, and assessing forces to impose the commander's will on a hostile, thinking, and adaptive enemy. Battle command applies leadership to translate decisions into actions—by synchronizing forces and warfighting functions in time, space, and purpose—to accomplish missions**. Battle command is guided by professional judgment gained from experience, knowledge, education, intelligence, and intuition. It is driven by commanders.

5-10.   Successful battle command demands timely and effective decisions based on applying judgment to available information. It requires knowing both when and what to decide. It also requires commanders to evaluate the quality of information and knowledge. Commanders identify important information requirements and focus subordinates and the staff on answering them. Commanders are aware that, once executed, the effects of their decisions are frequently irreversible. Therefore, they anticipate actions that follow their decisions.

5-11.   Commanders continuously combine analytic and intuitive approaches to decisionmaking to exercise battle command. Analytic decisionmaking approaches a problem systematically. The analytic approach aims to produce the *optimal* solution to a problem from among the solutions identified. The Army's analytic approach is the military decisonmaking process (MDMP). In contrast, *intuitive decisionmaking* is the act of reaching a conclusion that emphasizes pattern recognition based on knowledge, judgment, experience, education, intelligence, boldness, perception, and character. This approach focuses on assessment of the situation vice comparison of multiple options (FM 6-0). It relies on the experienced commander's and staff member's intuitive ability to recognize the key elements and implications of a particular problem or situation, reject the impractical, and select an *adequate* solution. (FM 5-0 discusses the MDMP. FM 6-0 discusses analytic and intuitive decisionmaking.)

5-12. The two approaches are not mutually exclusive. Commanders may make an intuitive decision based on situational understanding gained during the MDMP. If time permits, the staff may use a specific MDMP step, such as war-gaming, to validate or refine the commander's intuitive decision. When conducting the MDMP in a time-constrained environment, many techniques—such as selecting a single course of action—rely heavily on intuitive decisions. Even in the most rigorous, analytic decisonmaking processes, intuition sets boundaries for analysis.

5-13. Commanders understand, visualize, describe, direct, lead, and assess throughout the operations process. (See figure 5-1.) First, they develop a personal and in-depth understanding of the enemy and operational environment. Then they visualize the desired end state and a broad concept of how to shape the current conditions into the end state. Commanders describe their visualization through the commander's intent, planning guidance, and concept of operations in a way that brings clarity to an uncertain situation. They also express gaps in relevant information as commander's critical information requirements (CCIRs). Direction is implicit in command; commanders direct actions to achieve results and lead forces to mission accomplishment.

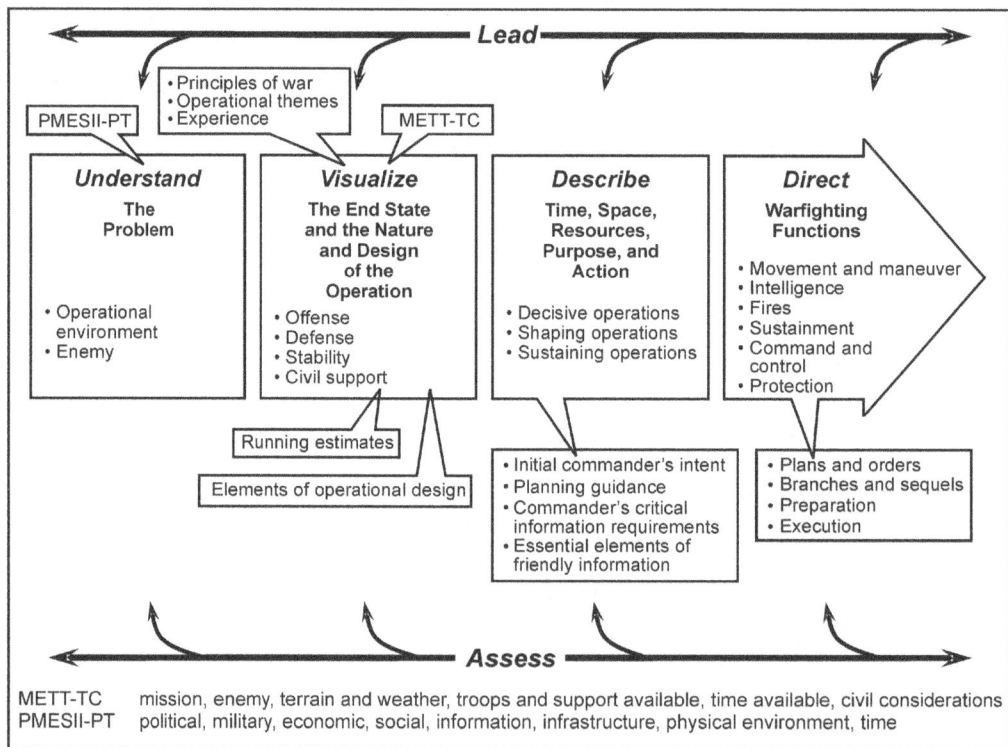

Figure 5-1. Battle command

5-14. Effective battle command requires commanders to continuously assess and lead. Assessment helps commanders better understand current conditions and broadly describe future conditions that define success. They identify the difference between the two and visualize a sequence of actions to link them. Commanders lead by force of example and personal presence. Leadership inspires Soldiers (and sometimes civilians) to accomplish things that they would otherwise avoid. This often requires risk. Commanders anticipate and accept risk to create opportunities to seize, retain, and exploit the initiative and achieve decisive results.

5-15. Battle command encourages the leadership and initiative of subordinates through mission command. Commanders accept setbacks that stem from the initiative of subordinates. They understand that land warfare is chaotic and unpredictable and that action is preferable to passivity. They encourage subordinates to accept calculated risks to create opportunities, while providing intent and control that allow for latitude and discretion.

UNDERSTAND

5-16. Understanding is fundamental to battle command. It is essential to the commander's ability to establish the situation's context. Analysis of the enemy and the operational variables provides the information senior commanders use to develop understanding and frame operational problems. (See paragraphs 1-22 through 1-40.) To develop a truer understanding of the operational environment, commanders need to circulate throughout their areas of operations as often as possible, talking to the subordinate commanders and Soldiers conducting operations, while observing for themselves. These individuals will have a more finely attuned sense of the local situation, and their intuition may cause them to detect trouble or opportunity long before the staff might. This deepens commanders' understanding. It allows them to anticipate potential opportunities and threats, information gaps, and capability shortfalls. Understanding becomes the basis of the commander's visualization.

5-17. Numerous factors determine the commander's depth of understanding. These include the commander's education, intellect, experience, and perception. Intelligence, surveillance, and reconnaissance (ISR) is indispensable as is actual observation and listening to subordinates. Formulating CCIRs, keeping them current, determining where to place key personnel, and arranging for liaisons also contribute to understanding. Maintaining understanding is a dynamic ability; a commander's situational understanding changes as an operation progresses. Relevant information fuels understanding and fosters initiative. Greater understanding enables commanders to make better decisions. It allows them to focus their intuition on visualizing the current and future conditions of the environment and describe them to subordinates.

VISUALIZE

5-18. Assignment of a mission provides the focus for developing the commander's visualization. Because military operations are fundamentally dynamic, this visualization must be continuous. Visualizing the desired end state requires commanders to clearly understand the operational environment and analyze the situation in terms of METT-TC. (See paragraph 5-24.) This analysis forms the basis of their situational understanding. Commanders consider the current situation and perform a mission analysis that assists in their initial visualization. They continually validate their visualization throughout the operation. To develop their visualization, commanders draw on several sources of knowledge and relevant information. These include—

- The elements of operational design appropriate to their echelon.
- Input from the staff and other commanders.
- Principles of war.
- Operational themes and related doctrine.
- Running estimates.
- The common operational picture.
- Their experience and judgment.
- Subject matter experts.

Visualization allows commanders to develop their intent and planning guidance for the entire operation, not just the initial onset of action.

Commander's Visualization

5-19. *Commander's visualization* **is the mental process of developing situational understanding, determining a desired end state, and envisioning the broad sequence of events by which the force will achieve that end state**. It involves discussion and debate between commanders and staffs. During planning, commander's visualization provides the basis for developing plans and orders. During execution, it helps commanders determine if, when, and what to decide as they adapt to changing conditions. Commanders and staffs continuously assess the progress of operations toward the desired end state. They plan to adjust operations as required to accomplish the mission.

5-20. Subordinate, supporting, adjacent, and higher commanders communicate with one another to compare perspectives and visualize their environment. Commanders increase the breadth and depth of their visualizations by collaborating with other commanders and developing a shared situational understanding. Likewise, staff input, in the form of running estimates, focuses analysis and detects potential effects on operations. Commanders direct staffs to provide the information necessary to shape their visualization.

5-21. Commanders consider the elements of operational design as they frame the problem and describe their visualization. (See chapter 6.) However, the utility and applicability of some elements are often limited at the tactical level. Commanders use the elements that apply to their echelon and situation.

Area of Influence

5-22. An *area of influence* is a geographical area wherein a commander is directly capable of influencing operations by maneuver and fire support systems normally under the commander's command or control (JP 1-02). The area of influence normally surrounds and includes the area of operations. (See paragraphs 5-77 through 5-79.) Understanding the command's area of influence helps the commander and staff plan branches to the current operation that could require the force to use capabilities outside the area of operations.

Area of Interest

5-23. An *area of interest* is that area of concern to the commander, including the area of influence, areas adjacent thereto, and extending into enemy territory to the objectives of current or planned operations. This area also includes areas occupied by enemy forces who could jeopardize the accomplishment of the mission (JP 2-03). The area of interest for stability or civil support operations may be much larger than that associated with offensive and defensive operations.

Mission Variables: The Factors of METT-TC

5-24. METT-TC is a memory aid that identifies the mission variables: **M**ission, **E**nemy, **T**errain and weather, **T**roops and support available, **T**ime available, and **C**ivil considerations. It is used in information management (the major categories of relevant information) and in tactics (the major variables considered during mission analysis). Mission analysis describes characteristics of the area of operations in terms of METT-TC, focusing on how they might affect the mission. (FM 6-0 discusses METT-TC in more detail.)

Mission

5-25. The *mission* is the task, together with the purpose, that clearly indicates the action to be taken and the reason therefor (JP 1-02). Commanders analyze a mission in terms of specified tasks, implied tasks, and the commander's intent two echelons up. They also consider the missions of adjacent units to understand their relative contributions to the decisive operation. Results of that analysis yield the essential tasks that—with the purpose of the operation—clearly specify the actions required. This analysis also produces the unit's mission statement—a short description of the task and purpose that clearly indicates the action to be taken and the reason for doing so. It contains the elements of who, what, when, where, and why. Mission command requires that commanders clearly communicate—and subordinates understand—the purpose for conducting an operation or a task.

5-26. When assigning missions, commanders ensure each subordinate's mission supports the decisive operation and the higher commander's intent. (See paragraphs 5-55 and 5-59 for definitions.) They identify the purpose for each task assigned, nesting unit missions with one another and with the decisive operation. (FM 5-0 discusses nesting.) Under mission command, commanders articulate each subordinate's mission in terms that foster the greatest possible freedom of action.

Enemy

5-27. The second variable to consider is the enemy. Relevant information regarding the enemy may include the following:

- Dispositions (including organization, strength, location, and mobility).
- Doctrine (or known execution patterns).
- Personal habits and idiosyncrasies.
- Equipment, capabilities, and vulnerabilities.
- Probable courses of action.

This analysis includes not only the known enemy but also other threats to mission success. These include threats posed by multiple adversaries with a wide array of political, economic, religious, and personal motivations.

5-28. To visualize threat capabilities and vulnerabilities, commanders require detailed, timely, and accurate intelligence. Of all relevant information, intelligence is the most uncertain. Commanders use ISR to collect the most important threat-related information and process it into intelligence. The intelligence officer synchronizes ISR. The operations officer integrates ISR through the operations process.

Terrain and Weather

5-29. Terrain and weather are natural conditions that profoundly influence operations. Terrain and weather are neutral; they favor neither side unless one is more familiar with—or better prepared to operate in—the physical environment. Terrain includes natural features (such as rivers and mountains) and man-made features (such as cities, airfields, and bridges). Terrain directly affects the selection of objectives and the location, movement, and control of forces. It also influences protective measures and the effectiveness of weapons and other systems. Effective use of terrain reduces the effects of enemy fires, increases the effects of friendly fires, and facilitates surprise. Terrain appreciation—the ability to predict its impact on operations—is an important skill for every leader. For tactical operations, terrain is analyzed using the five military aspects of terrain, expressed in the memory aid, OAKOC: **O**bservation and fields of fire, **A**venues of approach, **K**ey and decisive terrain, **O**bstacles, **C**over and concealment.

5-30. Climate and weather affect all operations. Climate is the prevailing pattern of temperature, wind velocity, and precipitation in a specific area measured over a period of years. Climate is a more predictable phenomenon than weather. It is also better suited to operational-level analysis. Planners typically focus analysis on how climate affects large-scale operations over a geographically diverse area. In contrast, weather describes the conditions of temperature, wind velocity, precipitation, and visibility at a specific place and time. It is more applicable to tactical analysis, where its effect on operations is limited in scale and duration. Climate and weather present opportunities and challenges in every operation. They affect conditions and capabilities of Soldiers and weapons systems, including mobility, obstacle emplacement times, and munitions performance. Effective commanders use climate and weather to their advantage.

Troops and Support Available

5-31. The fourth mission variable is the number, type, capabilities, and condition of available friendly troops and support. These include resources from joint, interagency, multinational, host-nation, commercial (via contracting), and private organizations. It also includes support provided by civilians. Commanders and staffs maintain information on friendly forces two echelons down. They track subordinate readiness—including training, maintenance, logistics, and morale. Commanders provide subordinates with the mix of troops and support needed to accomplish their missions. When assigning or allocating troops to subordinates, commanders consider differences in mobility, protection, firepower, equipment, morale, experience, leadership, and training.

5-32. Commanders consider available troops and support when determining the resources required to accomplish a mission—a troop-to-task analysis. If commanders determine they lack sufficient resources, they request additional support. When the resources needed to execute simultaneous operations are not available, commanders execute sequential operations.

Time Available

5-33. Time is critical to all operations. Controlling and exploiting it is central to initiative, tempo, and momentum. By exploiting time, commanders can exert constant pressure, control the relative speed of decisions and actions, and exhaust enemy forces. Upon receipt of a mission, commanders assess the time available for planning, preparing, and executing it. This includes the time required to assemble, deploy, and maneuver units to where they can best mass the effects of combat power. Commanders also consider how much time they can give subordinates to plan and prepare their own operations. Parallel and collaborative planning can help optimize available time. (FM 5-0 discusses parallel and collaborative planning.) At the operational level, planners consider longer spans of time from the friendly, enemy, and civilian perspectives.

5-34. Commanders also relate time to the enemy and conditions. As part of this analysis, commanders consider time in two contexts: First, they estimate the time available to friendly forces to accomplish the mission relative to enemy efforts to defeat them. Second, they consider the time needed to accomplish their objectives or to change current conditions into those of the desired end state. Analyzing the time available helps commanders determine how quickly and how far in advance to plan operations. The more time the commander and staff take, the more time the enemy has. The time spent perfecting a plan may work to the enemy's advantage; the additional time provided to enemy forces often offsets the minor gains a slightly improved plan gives friendly forces.

5-35. Commanders consider the time available relative to the situation. This analysis is essential when success depends on preventing the situation from deteriorating further. Ultimately, good plans executed sooner produce better results than perfect plans executed later.

Civil Considerations

5-36. Understanding the operational environment requires understanding civil considerations. *Civil considerations* reflect how the man-made infrastructure, civilian institutions, and attitudes and activities of the civilian leaders, populations, and organizations within an area of operations influence the conduct of military operations (FM 6-0). Commanders and staffs analyze civil considerations in terms of the categories expressed in the memory aid ASCOPE: **A**reas, **S**tructures, **C**apabilities, **O**rganizations, **P**eople, **E**vents.

5-37. Civil considerations help commanders develop an understanding of the social, political, and cultural variables within the area of operations and how these affect the mission. Understanding the relationship between military operations and civilians, culture, and society is critical to conducting full spectrum operations. (FM 3-05.40 contains additional information.) These considerations relate directly to the effects of the other instruments of national power. They provide a vital link between actions of forces interacting with the local populace and the desired end state.

5-38. Civil considerations are essential to developing effective plans for all operations—not just those dominated by stability or civil support. Full spectrum operations often involve stabilizing the situation, securing the peace, building host-nation capacity, and transitioning authority to civilian control. Combat operations directly affect the populace, infrastructure, and the force's ability to transition to host-nation authority. The degree to which the populace is expected to support or resist Army forces also affects the design of offensive and defensive operations.

5-39. Commanders use personal knowledge, area studies, and the intelligence and civil affairs running estimates to assess social, economic, and political factors. Commanders consider how these factors may relate to potential lawlessness, subversion, or insurgency. Their goal is to develop their understanding to the level of cultural astuteness. At this level, they can estimate the effects of friendly actions across the entire set of civil considerations and direct their subordinates with confidence. By increasing their knowledge of the human variables in the operational environment, commanders and staffs improve the force's ability to accomplish its missions. Cultural awareness improves how Soldiers interact with the populace and deters their false or unrealistic expectations. They have more knowledge of the society's common practices, perceptions, assumptions, customs, and values, giving better insight into the intent of individuals and groups.

DESCRIBE

5-40. After commanders visualize an operation, they describe it to their staffs and subordinates to facilitate shared understanding of the mission and intent. Commanders ensure subordinates understand the visualization well enough to begin planning. Commanders describe their visualization in doctrinal terms, refining and clarifying it as circumstances require. Commanders express their initial visualization in terms of—

- Initial commander's intent.
- Planning guidance, including an initial concept of operations.
- Information required for further planning (CCIRs).
- Essential elements of friendly information (EEFIs) that must be protected.

Initial Commander's Intent

5-41. Commanders summarize their visualization in their initial intent statement. The purpose of the initial commander's intent is to facilitate planning while focusing the overall operations process. Commanders develop this intent statement personally. It is a succinct description of the commander's visualization of the entire operation, a clear statement of what the commander wants to accomplish. The initial commander's intent links the operation's purpose with the conditions that define the desired end state. Usually the intent statement evolves as planning progresses and more information becomes available.

5-42. The initial commander's intent statement focuses the staff during the operations process. The staff uses this statement to develop and refine courses of action that contribute to establishing conditions that define the end state. Planning involves developing lines of effort that link the execution of tactical tasks to end state conditions. A clear initial intent statement is essential to this effort. (Paragraphs 6-66 through 6-71 discuss lines of effort.)

Planning Guidance

5-43. Commanders provide planning guidance with their initial intent statement. Planning guidance conveys the essence of the commander's visualization. Guidance may be broad or detailed, depending on the situation. Effective planning guidance is essentially an initial concept of operations that includes priorities for each warfighting function. It reflects how the commander sees the operation unfolding. It broadly describes when, where, and how the commander intends to employ combat power to accomplish the mission within the higher commander's intent.

5-44. Commanders use their experience and judgment to add depth and clarity to their planning guidance. They ensure staffs understand the broad outline of their visualization while allowing the latitude necessary to explore different options. This guidance provides the basis for a detailed concept of operations without dictating the specifics of the final plan. As with their intent, commanders may modify planning guidance based on staff and subordinate input and changing conditions.

Commander's Critical Information Requirements

5-45. A *commander's critical information requirement* is an information requirement identified by the commander as being critical to facilitating timely decisionmaking. The two key elements are friendly force information requirements and priority intelligence requirements (JP 3-0). A CCIR directly influences decisionmaking and facilitates the successful execution of military operations. Commanders decide whether to designate an information requirement as a CCIR based on likely decisions and their visualization of the course of the operation. A CCIR may support one or more decisions. During planning, staffs recommend information requirements for commanders to designate as CCIRs. During preparation and execution, they recommend changes to CCIRs based on assessment. A CCIR is—

- Specified by a commander for a specific operation.
- Applicable only to the commander who specifies it.
- Situation dependent—directly linked to a current or future mission.
- Focused on predictable events or activities.
- Time-sensitive—the answer to a CCIR must be reported to the commander immediately by any means available.
- Always promulgated by a plan or order.

5-46. Commanders limit the number of CCIRs to focus the efforts of limited collection assets. Typically, commanders identify ten or fewer CCIRs. The fewer the CCIRs, the easier it is for staffs to remember, recognize, and act on each one. This helps staffs and subordinates identify information the commander needs immediately. The staff's first priority is to provide the commander with answers to CCIRs. While most staffs provide relevant information, a good staff expertly distills that information. It identifies answers to CCIRs and gets them to the commander immediately. It also identifies vital information that does not answer a CCIR but that the commander nonetheless needs to know. A good staff develops this acumen through training and experience. Designating too many CCIRs limits the staff's ability to immediately recognize and react to them. Excessive critical items reduce the focus of collection efforts.

5-47. The list of CCIRs constantly changes. Commanders add and delete individual requirements throughout an operation based on the information needed for specific decisions. Commanders determine their own CCIRs but may select some from staff nominations. Staff sections recommend the most important priority intelligence requirements (PIRs) and friendly force information requirements (FFIRs) for the commander to designate as CCIRs. Once approved, a CCIR falls into one of two categories: PIRs and FFIRs.

5-48. A *priority intelligence requirement* is an intelligence requirement, stated as a priority for intelligence support, that the commander and staff need to understand the adversary or the operational environment (JP 2-0). PIRs identify the information about the enemy, terrain and weather, and civil considerations that the commander considers most important. Lessons from recent operations show that intelligence about civil considerations may be as critical as intelligence about the enemy. Thus, all staff sections may recommend information about civil considerations as PIRs. The intelligence officer manages PIRs for the commander.

5-49. A *friendly force information requirement* is information the commander and staff need to understand the status of friendly force and supporting capabilities (JP 3-0). FFIRs identify the information about the mission, troops and support available, and time available for friendly forces that the commander considers most important. In coordination with the staff, the operations officer manages FFIRs for the commander.

Essential Elements of Friendly Information

5-50. **An *essential element of friendly information* is a critical aspect of a friendly operation that, if known by the enemy, would subsequently compromise, lead to failure, or limit success of the operation, and therefore should be protected from enemy detection**. Although EEFIs are not CCIRs, they have the same priority. An EEFI establishes an element of information to protect rather than one to collect. EEFIs identify those elements of friendly force information that, if compromised, would jeopardize mission success.

5-51. EEFIs help commanders protect vital friendly information. Their identification is the first step in the operations security process and central to information protection. (FM 3-13 addresses the operations security process.) EEFIs are also key factors in formulating military deception operations. Operations security, information protection, and military deception all contribute to information superiority. (See chapter 7.)

DIRECT

5-52. Commanders direct all aspects of operations. This direction takes different forms during planning, preparation, and execution. Commanders make decisions and direct actions based on their situational understanding, which they maintain by continuous assessment. They use control measures to focus the operation on the desired end state. (Paragraph 5-72 defines control measure.) Commanders direct operations by—

- Preparing and approving plans and orders.
- Assigning and adjusting missions, tasks, task organization, and control measures based on changing conditions.
- Positioning units to maximize combat power, anticipate actions, or create or preserve maneuver options.
- Positioning key leaders to ensure observation and supervision at critical times and places.
- Adjusting support priorities and allocating resources based on opportunities and threats.
- Accepting risk to create opportunities to seize, retain, and exploit the initiative.
- Committing reserves.
- Changing support arrangements.

Plans and Orders

5-53. Plans and orders are key tools used by commanders in directing operations. Under mission command, commanders direct with mission orders. **Mission orders is a technique for developing orders that emphasizes to subordinates the results to be attained, not how they are to achieve them. It provides maximum freedom of action in determining how to best accomplish assigned missions**. Mission orders synchronize subordinates' actions only as required for mission success. Constraints are appropriate when mission success requires closely synchronized action by multiple elements. Even then, commanders establish constraints in a manner that least limits individual initiative.

5-54. When close coordination is necessary, commanders limit subordinates' freedom of action with control measures specified in plans and orders. Generally, subordinate commanders exercise full freedom of action within the concept of operations and commander's intent. Higher commanders may impose additional control over subordinates during a particular phase or mission. As soon as conditions allow, subordinates regain their freedom of action. Effective mission orders communicate to subordinates the situation, their commander's mission and intent, and the important tasks of each unit. The commander's intent and concept of operations set guidelines that ensure unity of effort while allowing subordinate commanders to exercise initiative.

Commander's Intent

5-55. **The commander's intent is a clear, concise statement of what the force must do and the conditions the force must establish with respect to the enemy, terrain, and civil considerations that represent the desired end state**. The commander's intent succinctly describes what constitutes success in an operation. It includes the operation's purpose and the conditions that define the end state. It links the mission, concept of operations, and tasks to subordinate units. A clear commander's intent facilitates a shared understanding and focus on the overall conditions that represent mission accomplishment. During execution, the commander's intent spurs individual initiative.

5-56. The commander's intent must be easy to remember and clearly understood two echelons down. The shorter the commander's intent, the better it serves these purposes. Typically, the commander's intent statement is three to five sentences long.

5-57. Commanders develop their intent statement personally. Commander's intent, coupled with mission, directs subordinates toward mission accomplishment, especially when current orders no longer fit the situation and subordinates must decide how to deviate from them. Subordinates use the commander's intent to orient their efforts and help them make decisions when facing unforeseen opportunities or threats.

Concept of Operations

5-58. The *concept of operations* **is a statement that directs the manner in which subordinate units cooperate to accomplish the mission and establishes the sequence of actions the force will use to achieve the end state. It is normally expressed in terms of decisive, shaping, and sustaining operations.** The concept of operations expands on the commander's intent by describing how the commander wants the force to accomplish the mission. It states the principal tasks required, the responsible subordinate units, and how the principal tasks complement one another. The concept of operations promotes general understanding by stating the task (such as attack) that directly accomplishes the mission (the decisive operation) and the units that will execute it. The concept of operations clearly describes the other units' tasks in terms of shaping and sustaining operations. It may include, for example, the type of support or specific location for providing support. Normally, the concept of operations projects the status of the force at the end of the operation. If the mission dictates a significant change in tasks during the operation, the commander may phase the operation. Small-unit commanders and leaders usually do not describe their concept of operations in terms of decisive, shaping, and sustaining operations; they simply assign tasks to subordinates using main effort as required. (FM 5-0 discusses the concept of operations in detail.)

Decisive Operations

5-59. The *decisive operation* **is the operation that directly accomplishes the mission. It determines the outcome of a major operation, battle, or engagement. The decisive operation is the focal point around which commanders design the entire operation.** Multiple units may be engaged in the same decisive operation. For example, one task force may follow another on an axis of advance, prepared to assume the attack. Units operating in noncontiguous areas of operations may execute the tasks composing the higher headquarters' decisive operation simultaneously in different locations. Commanders visualize the decisive operation and then design shaping and sustaining operations around it.

5-60. Changing the mission normally changes the decisive operation. This can occur because of a change of phase but is more typical in the conduct of branches and sequels. When Army forces transition to a new operation, through either mission accomplishment or a significant change in the situation, the commander identifies a new decisive operation. Any element of full spectrum operations can be the decisive operation, as can a specific task, such as movement to contact or civil security. The commander decides. In a protracted stability operation, for example, the commander may identify a stability task in a particular area as the decisive operation: "The decisive operation is to provide civil security in Tal Afar. Task Force Roper provides civil security in Tal Afar commencing D+2."

Shaping Operations

5-61. A *shaping operation* **is an operation at any echelon that creates and preserves conditions for the success of the decisive operation.** Shaping operations establish conditions for the decisive operation through effects on the enemy, population (including local leaders), and terrain. Information engagement, for example, may reduce tensions between Army units and different ethnic groups through direct contact between Army leaders and local leaders. Shaping operations may occur throughout the operational area and involve any combination of forces and capabilities.

5-62. Shaping operations may occur before, during, or after the decisive operation begins. Some shaping operations, especially those executed simultaneously with the decisive operation, may be economy of force actions. However, if the force available does not permit simultaneous decisive and shaping operations, the commander sequences shaping operations around the decisive operation. The concept of operations describes how shaping operations contribute to the decisive operation's success, often in terms of the purpose. For example, "Task Force Hammer conducts search and attack operations in area of operations Anvil to neutralize insurgents that threaten Tal Afar. Task Force Rapier secures area of operations Sparrow as a support area for brigade operations."

Sustaining Operations

5-63. **A *sustaining operation* is an operation at any echelon that enables the decisive operation or shaping operations by generating and maintaining combat power**. Sustaining operations differ from decisive and shaping operations in that they are focused internally (on friendly forces) rather than externally (on the enemy or environment). They typically address important sustainment and protection actions essential to the success of decisive and shaping operations. However, sustaining operations cannot be decisive themselves. Note that logistic and medical support provided to the civilian population are tasks related to stability and civil support operations (provide essential services); they are not sustaining operations. At the operational level, sustaining operations focus on preparing the force for the operation's next phase. They determine the limit of operational reach. At the tactical level, sustaining operations determine the tempo of the overall operation; they ensure the force is able to seize, retain, and exploit the initiative.

5-64. Sustaining operations are continuous; commanders do not reiterate routine sustainment requirements in the concept of operations. Rather, the concept of operations emphasizes important changes in sustainment required by the operation. For example, "Brigade support battalion moves to area of operations Sparrow as soon as it is secure and establishes the brigade support area. On order, brigade special troops battalion assumes control of area of operations Sparrow." If there are no significant changes to sustainment, it is discussed in the sustainment paragraph or annexes.

Main Effort

5-65. The concept of operations identifies a main effort unit if required; otherwise, the priorities of support go to the unit conducting the decisive operation. **The *main effort* is the designated subordinate unit whose mission at a given point in time is most critical to overall mission success. It is usually weighted with the preponderance of combat power**. Designating a main effort temporarily prioritizes resource allocation. When commanders designate a unit as the main effort, it receives priority of support and resources. Commanders shift resources and priorities to the main effort as circumstances and the commander's intent require. Commanders may shift the main effort several times during an operation. A unit conducting a shaping operation may be designated as the main effort until the decisive operation commences. However, the unit with primary responsibility for the decisive operation becomes the main effort upon execution of the decisive operation. For example, "Task Force Hammer is the main effort until D+2."

LEAD

5-66. After commanders make decisions, they guide their forces throughout execution. During execution, commanders must provide the strength of character, moral courage, and will to follow through with their decisions. When changing decisions, they must know when and what to decide, and when to make other decisions that address changes in the situation. (FM 6-22 discusses leadership actions during execution.)

5-67. In many instances, a leader's physical presence is necessary to lead effectively. Advanced information systems provide detailed information that facilitates situational understanding and command and control; however, much of the art of command stems from intuition. Commanders carefully consider where they need to be, balancing the need to inspire Soldiers with that of maintaining an overall perspective of the entire operation. The commander's forward presence demonstrates a willingness to share danger. It also allows them to appraise for themselves the subordinate unit's condition, including leader and Soldier morale. Forward presence allows commanders to sense the human dimension of conflict, particularly when fear and fatigue reduce effectiveness. Then commanders need to lead by example, face-to-face with Soldiers. Commanders cannot let the perceived advantages of improved information technology compromise their obligation to lead by example.

5-68. The commander's will is the one constant element that propels the force through the shock and friction of battle. Friction is inherent in all operations—inevitably, things can and will go wrong. The ability of leaders and Soldiers to concentrate erodes as they reach the limit of their endurance. Against a skilled and resolute enemy, Soldiers may approach that point when fear, uncertainty, and physical exhaustion dominate their thinking. It is then that the commander's strength of will and personal presence provide the moral impetus for actions that lead to victory.

CONTROL

5-69. Control allows commanders to adjust operations to conform to their commander's intent as conditions change. Staffs provide their greatest support in assisting the commander with control. Commanders use two basic forms of control—procedural and positive. Commanders balance the two based on the situation:

- *Army procedural control* **is a technique of regulating forces that relies on a combination of orders, regulations, policies, and doctrine (including tactics, techniques, and procedures).** Army procedural control requires no intervention by the higher headquarters once it is established.

- *Army positive control* **is a technique of regulating forces that involves commanders and leaders actively assessing, deciding, and directing them.** It may be restrictive in that commanders directly monitor operations and intervene to direct specific actions to better synchronize subordinates' operations. Excessive use of Army positive control can rapidly become detailed command. (FM 6-0 describes detailed command.)

5-70. The definitions of individual control measures provide Army procedural control without requiring detailed explanations. For example, boundaries, the most important control measure, designate the area of operations assigned to or by a commander. Commanders know they have full freedom of action within their area of operations.

5-71. Army positive control may be appropriate when detailed direction is needed to sequence actions or coordinate the activities of forces close to each other. An example of such a situation is a river crossing. In exercising Army positive control, commanders may use digital information systems to assess without requesting information explicitly or continuously from subordinates. Positive control supplements mission command when necessary.

CONTROL MEASURES

5-72. Commanders exercise Army procedural and Army positive control through control measures. **A** *control measure* **is a means of regulating forces or warfighting functions.** Control measures are established under a commander's authority; however, commanders may authorize staff officers and subordinate leaders to establish them. Commanders may use control measures for several purposes: for example, to assign responsibilities, require synchronization between forces, impose restrictions, or establish guidelines to regulate freedom of action. Control measures are essential to coordinating subordinates' actions. They can be permissive or restrictive. Permissive control measures allow specific actions to occur; restrictive control measures limit the conduct of certain actions.

5-73. Control measures help commanders direct by establishing responsibilities and limits that prevent subordinate units' actions from impeding one another. They foster coordination and cooperation between forces without unnecessarily restricting freedom of action. Good control measures foster freedom of action, decisionmaking, and individual initiative.

5-74. Control measures may be detailed (such as a division operation order) or simple (such as a checkpoint). Control measures include but are not limited to—

- Laws and regulations.
- Planning guidance.
- Delegation of authority.
- Specific instructions to plans and orders and their elements, including—
  - Commander's intent.
  - Unit missions and tasks.
  - CCIRs.
  - EEFIs.
  - Task organization.
  - Concept of operations.

- ■ Target lists.
- ■ Rules of engagement.
- ■ Service support plans.
- ■ Graphic control measures.
- ■ Unit standing operating procedures that control actions as reporting and battle rhythm.
- ● Information requirements.

5-75. Certain control measures belong to the commander alone and may not be delegated. These include the commander's intent, unit mission statement, planning guidance, CCIRs, and EEFIs. Unit standing operating procedures specify many control measures. An operation plan or order modifies them and adds additional measures for a specific operation. Commanders, assisted by their staffs, modify control measures to account for the dynamic conditions of operations.

5-76. Some control measures are graphic. **A *graphic control measure* is a symbol used on maps and displays to regulate forces and warfighting functions**. Graphic control measures are always prescriptive. They include symbols for boundaries, fire support coordination measures, some airspace control measures, air defense areas, and minefields. Commanders establish them to regulate maneuver, movement, airspace use, fires, and other aspects of operations. In general, all graphic control measures should relate to easily identifiable natural or man-made terrain features. (FM 1-02 portrays and defines graphic control measures and discusses rules for selecting and applying them.)

AREA OF OPERATIONS

5-77. One of the most basic and important control measures is the area of operations. The Army or land force commander is the supported commander within that area of operations designated by the joint force commander for land operations. Within their areas of operations, commanders integrate and synchronize maneuver, fires, and interdiction. To facilitate this integration and synchronization, commanders have the authority to designate targeting priorities and timing of fires within their areas of operations. Commanders consider a unit's area of influence when assigning it an area of operations. An area of operations should not be substantially larger than the unit's area of influence. Ideally, the entire area of operations is encompassed by the area of influence. An area of operations that is too large for a unit to control can allow sanctuaries for enemy forces and may limit joint flexibility.

5-78. Assigning areas of operations to subordinate commanders maximizes decentralized execution by empowering those commanders to exercise initiative. Mission command gives commanders the authority to create any effects necessary to accomplish the mission (consistent with the rules of engagement) within their areas of operations. However, commanders cannot create effects outside their areas of operations without permission from the commander assigned the area of operations in which those effects will occur. Further, commanders must control all parts of their area of operations not assigned to subordinates. Assignment of an area of operations includes authority to perform the following:

- ● Terrain management.
- ● Intelligence collection.
- ● Civil affairs activities.
- ● Air and ground movement control.
- ● Clearance of fires.
- ● Security.

5-79. Subordinate unit areas of operations may be contiguous or noncontiguous. (See figure 5-2.) A common boundary separates contiguous areas of operations. Noncontiguous areas of operations do not share a common boundary; the concept of operations provides procedural control of elements of the force. **An *unassigned area* is the area between noncontiguous areas of operations or beyond contiguous areas of operations. The higher headquarters is responsible for controlling unassigned areas within its area of operations.** Designating an unassigned area only indicates that the area is not assigned to a subordinate. Unassigned areas remain the responsibility of the controlling headquarters.

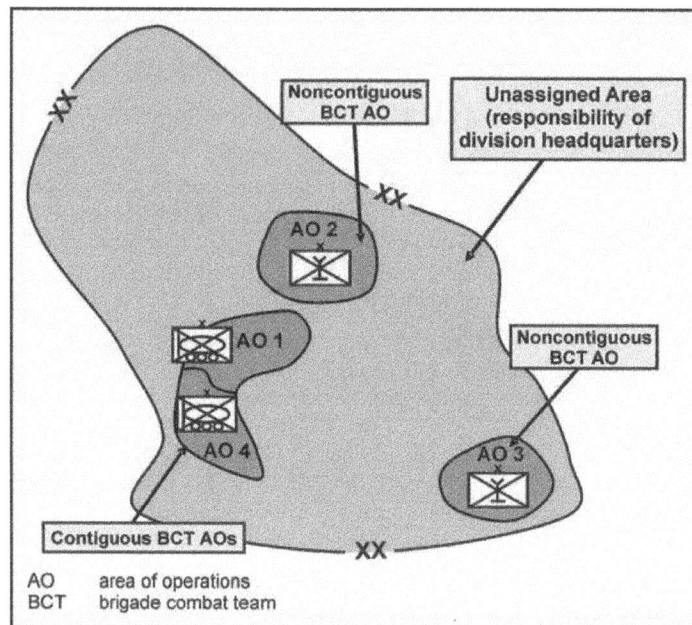

Figure 5-2. Contiguous, noncontiguous, and unassigned areas

COMMON OPERATIONAL PICTURE

5-80. The *common operational picture* is a single display of relevant information within a commander's area of interest tailored to the user's requirements and based on common data and information shared by more than one command. The availability of a common operational picture facilitates mission command. The common operational picture lets subordinates see the overall operation and their contributions to it as the operation progresses. This knowledge reduces the level of control higher commanders need to exercise over subordinates. The common operational picture features a scale and level of detail that meets the information needs of that commander and staff. It varies among staff sections and echelons. Separate echelons create a common operational picture by collaborating, sharing, and refining relevant information. To the extent permitted by technology, the common operational picture incorporates as much information from running estimates as possible.

THE OPERATIONS PROCESS

5-81. While differing significantly in design and application, all operations follow the operations process. **The *operations process* consists of the major command and control activities performed during operations: planning, preparing, executing, and continuously assessing the operation. The commander drives the operations process.** (See figure 5-3, page 5-16.) These activities may be sequential or simultaneous. They are usually not discrete; they overlap and recur as circumstances demand. Commanders use the operations process to help them decide when and where to make decisions, control operations, and provide command presence.

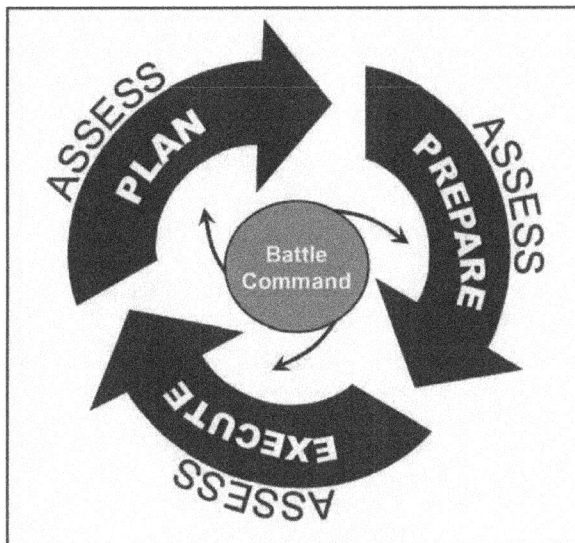

Figure 5-3. The operations process

5-82. Throughout the operations process, commanders synchronize forces and warfighting functions to accomplish missions. This synchronization is essential to achieving synergistic effects. It may be needed to achieve any effects at all. For example, delivery of fires must be synchronized with target acquisition to produce the desired effects. However, synchronization is not an end in itself. It is useful only as it contributes to the greater effectiveness of the force. Unnecessary synchronization or synchronization for limited gains degrades tempo, impedes initiative, and allows the enemy to act within the friendly force decision cycle. Excessive synchronization undermines mission command.

5-83. Commanders and staffs use the MDMP and troop leading procedures to integrate activities during planning. They also use several other processes and activities to synchronize operations throughout the operations process. Some occur only during preparation or execution; these are listed in the discussions of those activities. Others, the integrating processes and continuing activities, occur during all operations process activities. Paragraphs 5-113 through 5-115 address these processes and activities.

ASSESS

5-84. *Assessment* **is the continuous monitoring and evaluation of the current situation, particularly the enemy, and progress of an operation**. Commanders, assisted by their staffs and subordinate commanders, continuously assess the current situation and the progress of the operation and compare it with the concept of operations, mission, and commander's intent. Based on their assessment commanders direct adjustments, ensuring that the operation remains focused on the mission and commander's intent.

5-85. Assessment precedes and guides every operations process activity and concludes each operation or phase of an operation. It involves a comparison of forecasted outcomes to actual events. Assessment entails three tasks:

- Continuously assessing the enemy's reactions and vulnerabilities.
- Continuously monitoring the situation and progress of the operation towards the commander's desired end state.
- Evaluating the operation against measures of effectiveness and measures of performance.

5-86. A *measure of performance* is a criterion used to assess friendly actions that is tied to measuring task accomplishment (JP 3-0). Measures of performance answer the question, "Was the task or action performed as the commander intended?" A measure of performance confirms or denies that a task has been properly performed.

5-87. A *measure of effectiveness* is a criterion used to assess changes in system behavior, capability, or operational environment that is tied to measuring the attainment of an end state, achievement of an objective, or creation of an effect (JP 3-0). Measures of effectiveness focus on the results or consequences of actions taken. They answer the question, "Is the force doing the right things, or are additional or alternative actions required?" A measure of effectiveness provides a benchmark against which the commander assesses progress toward accomplishing the mission.

5-88. Commanders monitor the current situation for unanticipated successes, failures, or enemy actions. As commanders assess the operation, they look for opportunities, threats, and acceptable progress. They accept risks, seize opportunities, and mitigate threats. Throughout the operation, commanders visualize, describe, and direct changes to the operation.

5-89. Staffs analyze the current situation using mission variables and prepare their running estimates. **A *running estimate* is a staff section's continuous assessment of current and future operations to determine if the current operation is proceeding according to the commander's intent and if future operations are supportable.** Staffs continuously assess the impact of new information on the conduct of the operation. They update running estimates and determine if adjustments to the operation are required. Commanders empower their staffs to make adjustments within their areas of expertise. This requires staffs to understand the aspects of operations that require the commander's attention as opposed to those delegated to their control.

5-90. Commanders integrate their own assessments and those of subordinate commanders into all aspects of the operations process. Assessment helps commanders refine their situational understanding. It allows them to make informed, rational decisions throughout the entire operation. During planning, assessment focuses on understanding the current conditions in the operational environment and developing relevant courses of action. During preparation and execution, it emphasizes evaluating progress toward the desired end state, determining variances from expectations, and determining the significance (threat or opportunity) of those variances.

5-91. The common operational picture, observations of commanders, and running estimates are the primary tools for assessing the operation against the concept of operations, mission, and commander's intent. The commander's visualization forms the basis of the commander's personal decisonmaking methodology throughout the operation. Running estimates provide information, conclusions, and recommendations from the perspective of each staff section. They help to refine the common operational picture and supplement it with information not readily displayed.

5-92. Commanders avoid excessive analysis when assessing operations. Committing valuable time and energy to developing elaborate and time-consuming assessments squanders resources better devoted to other operations process activities. Effective commanders avoid burdening subordinates and staffs with overly detailed assessment and collection tasks. Generally, the echelon at which a specific operation, task, or action is conducted should be the echelon at which it is assessed. This provides a focus for assessment at each echelon. It enhances the efficiency of the overall operations process.

PLAN

5-93. A *plan* is a design for a future or anticipated operation (FM 5-0). This continuous, evolving framework of anticipated actions guides subordinates through each phase of the operation. A plan is a framework from which to adapt rather than a script to follow. The measure of a good plan is not whether execution transpires as planned but whether the plan facilitates effective action during unforeseen events. Good plans foster initiative, account for uncertainty and friction, and mitigate threats.

Army Planning

5-94. *Planning* **is the process by which commanders (and the staff, if available) translate the commander's visualization into a specific course of action for preparation and execution, focusing on the**

expected results. Planning begins with analysis and assessment of the conditions in the operational environment, with particular emphasis on the enemy, to determine the relationships among the mission variables. It involves understanding and framing the problem and envisioning the set of conditions that represent the desired end state. Based on the commander's guidance, planning includes formulating one or more supportable courses of action to accomplish the mission.

5-95. Commanders and staffs consider the consequences and implications of each course of action. Once the commander selects a course of action, the staff formulates specified tasks to subordinates, required staff actions, and an assessment framework. Planning develops the detailed information required during execution. Examples include setting initial conditions, assigning command relationships, and establishing priorities. Planning does not cease with production of a plan or order. It continues throughout an operation, as the order is refined based on changes in the situation. In addition, staffs refine plans for branches and sequels during an operation.

5-96. Whenever possible, commanders employ red teams to examine plans from an opponent's perspective. Red teams provide insight into possible flaws in the plan as well as potential reactions by the enemy and other people in the area of operations. This information helps the staff improve the plan and develop more effective branches and sequels.

5-97. The scope, complexity, and length of planning horizons differ at the operational and tactical levels. At the operational level, campaign planning coordinates major actions across significant periods. Planners integrate Service capabilities with those of joint, interagency, intergovernmental, and multinational organizations. (JP 5-0 contains doctrine for joint operation planning.) Tactical planning has the same clarity of purpose but typically reflects a shorter planning horizon. Comprehensive, continuous, and adaptive planning characterizes successful operations at both levels.

5-98. Army commanders use two doctrinal planning procedures to integrate activities during planning: the MDMP and troop leading procedures. In units with a formally organized staff, the MDMP provides structure to help commanders and staffs develop estimates, plans, and orders. It provides a logical sequence for decisionmaking and interaction between the commander and staff, and it provides a common framework for parallel planning. At the lowest tactical echelons, commanders and leaders follow troop leading procedures. Both the MDMP and troop leading procedures hinge on the commander's ability to visualize and describe the operation. They are means to an end; their inherent value lies in the results achieved, not the process. (FM 5-0 discusses the MDMP and troop leading procedures.)

5-99. Planning continues as necessary during preparation and execution. When circumstances are not suited for the MDMP or troop leading procedures, commanders rely on intuitive decisionmaking and direct contact with subordinate commanders to integrate activities.

Joint Planning

5-100. Joint interdependence requires brigade and higher headquarters to understand joint planning doctrine. Corps and theater army headquarters must be prepared to serve as the Army component of a joint force. These headquarters also serve as the base for joint task force headquarters. Since organizations filling these roles participate in joint planning and assessment, Army commanders and staffs must understand joint doctrine pertaining to effects.

5-101. Regardless of the role of effects in joint planning and assessment, joint force commanders issue orders to Service and functional component headquarters in the five-paragraph field order format. These orders assign tasks to subordinate units, detailing the effects to be achieved. For Army forces, this represents no change. Mission command and mission orders similarly focus on the effects to achieve rather than how to achieve them. Despite different terminology and processes, the use of effects in joint planning serves only to reinforce the essence of mission command: trust, initiative, and flexibility.

PREPARE

5-102. *Preparation* **consists of activities performed by units to improve their ability to execute an operation. Preparation includes, but is not limited to, plan refinement; rehearsals; intelligence,**

surveillance, and reconnaissance; coordination; inspections; and movement. Preparation creates conditions that improve friendly forces' opportunities for success. It facilitates and sustains transitions, including those to branches and sequels.

5-103.   Preparation requires staff, unit, and Soldier actions. Mission success depends as much on preparation as planning. Rehearsals help staffs, units, and Soldiers to better understand their roles in upcoming operations, practice complicated tasks, and ensure equipment and weapons function properly. Activities specific to preparation include—

- Revising and refining the plan.
- Rehearsals.
- Force tailoring and task-organizing.
- Surveillance and reconnaissance.
- Training.
- Troop movements.
- Precombat checks and inspections.
- Sustainment preparations.
- Integrating new Soldiers and units.
- Subordinate confirmation briefs and backbriefs.

5-104.   Several preparation activities begin during planning and continue throughout execution. For example, uncommitted forces prepare for contingencies identified in branches and subsequent events detailed in sequels. Committed units revert to preparation when they reach their objectives, occupy defensive positions, or pass into reserve.

EXECUTE

5-105.   *Execution* **is putting a plan into action by applying combat power to accomplish the mission and using situational understanding to assess progress and make execution and adjustment decisions**. It focuses on concerted action to seize, retain, and exploit the initiative. Army forces seize the initiative immediately and dictate tempo throughout all operations.

5-106.   Commanders use mission command to achieve maximum flexibility and foster individual initiative. Subordinates exercising their initiative can significantly increase the tempo of operations; however, this may desynchronize the overall operation. Desynchronization may reduce commanders' abilities to mass the effects of combat power. Executing even relatively minor, planned actions produces second- and third-order effects throughout the force; these affect the operation's overall synchronization. Nonetheless, under mission command, commanders accept some risk of desynchronization as the price of seizing, retaining, and exploiting the initiative.

5-107.   The commander's intent and mission orders focus every echelon on executing the concept of operations. Collaborative synchronization among subordinates is enabled and expected under mission command. It depends on individual initiative to resynchronize the overall operation continuously when subordinates exploit opportunities. Subordinates' successes may offer opportunities anticipated in the concept of operations or develop advantages that make a new concept practical. In either case, the commander's intent keeps the force acceptably focused and synchronized. Subordinates need not wait for top-down synchronization. The climate fostered by mission command encourages subordinates to act on information about the enemy, adversaries, events, and trends without detailed direction.

5-108.   Commanders assess the situation throughout execution based on information from the common operational picture, running estimates, and subordinate commanders' assessments. When the situation varies from the assumptions on which the order was based, commanders direct adjustments to exploit opportunities and counter threats. Possible adjustments include initiating a new phase of the operation or modifying the concept of operations.

5-109. Staffs help commanders control the integrating processes and continuing activities during execution. In addition, commanders, assisted by their staffs, perform the following activities specific to execution:

- Focus assets on the decisive operation.
- Adjust CCIRs and EEFIs based on the situation.
- Adjust control measures.
- Manage movement and positioning of supporting units.
- Adjust unit missions and tasks as necessary.
- Modify the concept of operations as required.
- Position or relocate committed, supporting, and reserve units.

5-110. As commanders assess the operation, they determine when decisions are required. Orders usually identify some decision points; however, unanticipated enemy actions or conditions often present situations that require unanticipated decisions. Commanders act when decisions are required. They do not wait for a set time in the battle rhythm.

5-111. During execution, commanders rely more on intuitive than on analytic decisionmaking. Commanders draw on experience, intellect, creativity, intuition, and education to make rapid decisions. They learn deliberately as the situation develops and make changes based on that learning. Staffs must synchronize or resynchronize forces and warfighting functions faster during execution than during planning and preparation. They must do this while forces are moving and processes are ongoing.

5-112. Commanders do not restrict their visualization to the current operation. As their situational understanding evolves, commanders incorporate considerations for the operation's next phase or sequel into their visualization. They begin to visualize how to transition from the current operation to the next one. Based on their visualization, commanders direct actions to posture the force for the transition. As they visualize the implications of events and their solutions, commanders describe their conclusions to staff and subordinates through updated CCIRs and planning guidance. The guidance may be to develop a branch or change the main effort to exploit success. Commanders direct adjustments when necessary, primarily through fragmentary orders.

INTEGRATING PROCESSES AND CONTINUING ACTIVITIES

5-113. Commanders use integrating processes and continuing activities to synchronize operations during all operations process activities. Figure 5-4 illustrates how the integrating processes and continuing activities last throughout the operations process. Battle rhythm is a key control measure for managing integration.

Integrating Processes

5-114. Certain integrating processes occur during all operations process activities. They must be synchronized with each other as well as integrated into the overall operation:

- Intelligence preparation of the battlefield. (For joint and functional component commanders, this is intelligence preparation of the operational environment. See JP 2-0.)
- Targeting. (See FM 6-20-10.)
- ISR synchronization.
- Composite risk management. (See FM 5-19.)
- Knowledge management.

Plan	Prepare	Execute
Assess		

Integrating Processes ————————▶
  • Intelligence preparation of the battlefield
  • Targeting
  • Intelligence, surveillance, and reconnaissance synchronization
  • Composite risk management
  • Knowledge management

Continuing Activities ————————▶
  • Intelligence, surveillance, and reconnaissance
  • Security operations
  • Protection
  • Liaison and coordination
  • Terrain management
  • Information management
  • Airspace command and control

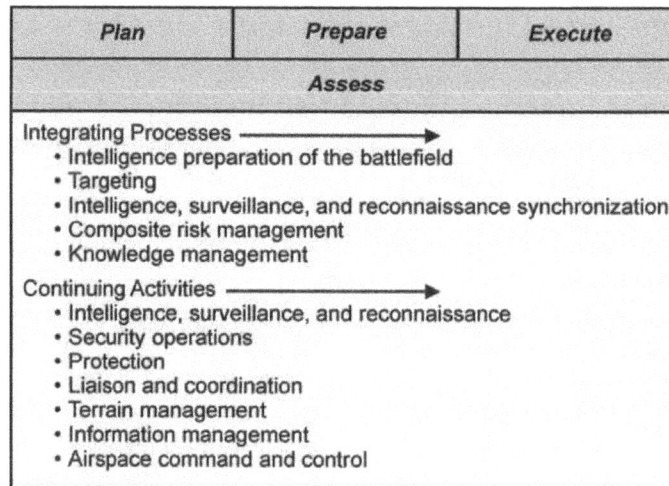

Figure 5-4. Operations process expanded

Continuing Activities

5-115.   The following activities continue during all operations process activities. They are synchronized with one another and integrated into the overall operation:

- Intelligence, surveillance, and reconnaissance.
- Security operations.
- Protection.
- Liaison and coordination.
- Terrain management.
- Information management.
- Airspace command and control.

SUMMARY

5-116.   Commanders execute command and control through battle command. They understand, visualize, describe, direct, lead, and assess operations in complex, dynamic environments. They combine the art of command and science of control throughout the operations process. Mission command is the preferred means of battle command. Commanders encourage individual initiative through mission orders and a climate of mutual trust and understanding. Guided by their experience, knowledge, education, intelligence, and intuition, commanders apply leadership to translate decisions into action. They exercise battle command in synchronizing forces and capabilities in time, space, and purpose to accomplish missions.

This page intentionally left blank.

Chapter 6

Operational Art

This chapter discusses operational art, including operational design and the levels of war. Operational art represents the creative aspect of operational-level command. It is the expression of informed vision across the levels of war. Operational design is a bridge between the strategic end state and the execution of tactical tasks. The elements of operational design help operational commanders clarify and refine their concept of operations by providing a framework to describe operations.

> *The first, the supreme, the most far-reaching act of judgment that the statesman and commander have to make is to establish...the kind of war on which they are embarking; neither mistaking it for, nor trying to turn it into, something that is alien to its nature. This is the first of all strategic questions and the most comprehensive.*

> Carl von Clausewitz
> *On War*[3]

UNDERSTANDING OPERATIONAL ART

6-1.   Military operations require integrating creative vision across the levels of war. Military art pervades operations at all echelons. Although military art transcends the levels of war, operational art is distinct. *Operational art* is the application of creative imagination by commanders and staffs—supported by their skill, knowledge, and experience—to design strategies, campaigns, and major operations and organize and employ military forces. Operational art integrates ends, ways, and means across the levels of war (JP 3-0). It is applied only at the operational level.

6-2.   Operational art reflects an intuitive understanding of the operational environment and the approach necessary to establish conditions for lasting success. In visualizing a campaign or major operation, operational commanders determine which conditions satisfy the President's strategic guidance; taken together, these conditions become the end state. Commanders devise and execute plans that complement the actions of the other instruments of national power in a focused, unified effort. To this end, operational commanders draw on experience, knowledge, education, intellect, intuition, and creativity.

THE LEVELS OF WAR

6-3.   The levels of war define and clarify the relationship between strategy, operational approach, and tactical actions. (See figure 6-1, page 6-2.) The levels have no finite limits or boundaries. They correlate to specific levels of responsibility and planning. They help organize thought and approaches to a problem. The levels clearly distinguish between headquarters and the specific responsibilities and actions performed at each echelon. Despite advances in technology, digital information sharing, and the increased visibility of tactical actions, the levels of war remain useful. Decisions at one level always affect other levels.

6-4.   A natural tension exists between the levels of war and echelons of command. This tension stems from different perspectives, requirements, and constraints associated with command at each level of war. Between the levels of war, the horizons for planning, preparation, and execution differ greatly. Operational-level commanders typically orchestrate the activities of military and other governmental

---

[3] © 1984. Reproduced with permission of Princeton University Press.

organizations across large areas. Tactical commanders focus primarily on employing combined arms within an area of operations. They may work with civilian agencies, but political, informational, and economic issues may not be as important at the tactical level. In many situations, tactical commanders receive missions that divert combat power from tasks that seem more urgent at lower levels. It is a commander's responsibility to recognize and resolve this tension.

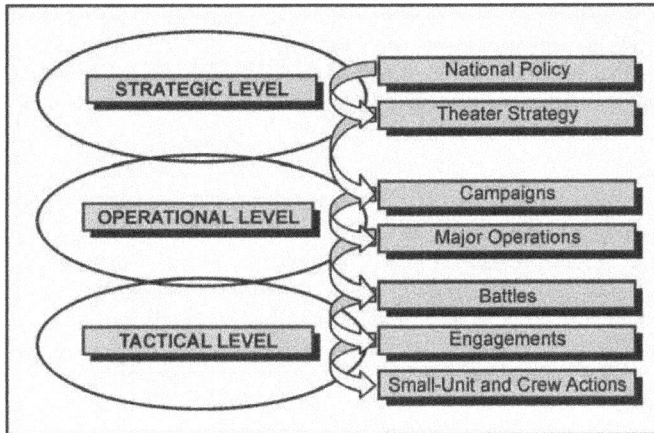

Figure 6-1. Levels of war

6-5.   A string of tactical victories does not guarantee success at the operational and strategic levels. Tactical success, while required to set operational conditions, must be tied to attaining the strategic end state. Wars are won at the operational and strategic levels; yet without tactical success, a major operation cannot achieve the desired end state. Commanders overcome this tension through open and continuous dialog, a thorough understanding of the situation across the levels of war, and a shared vision that integrates and synchronizes actions among the echelons of command.

6-6.   Individuals, crews, and small units act at the tactical level. At times, their actions may produce strategic or operational effects. However, this does not mean these elements are acting at the strategic or operational level. Actions are not strategic unless they contribute directly to achieving the strategic end state. Similarly, actions are considered operational only if they are directly related to operational movement or the sequencing of battles and engagements. The level at which an action occurs is determined by the perspective of the echelon in terms of planning, preparation, and execution.

STRATEGIC

6-7.   The *strategic level of war* is the level of war at which a nation, often as a member of a group of nations, determines national or multinational (alliance or coalition) strategic security objectives and guidance, and develops and uses national resources to achieve these objectives. Activities at this level establish national and multinational military objectives; sequence initiatives; define limits and assess risks for the use of military and other instruments of national power; develop global plans or theater war plans to achieve those objectives; and provide military forces and other capabilities in accordance with strategic plans (JP 3-0).

6-8.   *Strategy* is a prudent idea or set of ideas for employing the instruments of national power in a synchronized and integrated fashion to achieve theater, national, and/or multinational objectives (JP 3-0). The President translates national interests and policy into a national strategic end state. Combatant commanders base their theater strategic planning on this end state. To ensure their military strategy is consistent with national interests and policy, combatant commanders participate in strategic discourse with the President, Secretary of Defense (through the Chairman of the Joint Chiefs of Staff), and multinational partners.

Peacetime military engagement is vital to U.S. strategy and integral to theater security cooperation plans. Strategy involves more than campaigns and major operations. When successful, these plans promote national or multinational goals through peaceful processes. Peacetime military engagement contributes to the ability of multinational forces to operate together.

6-9. National interests and policy define and inform military strategy. They provide a broad framework for conducting operations. A combatant commander's military strategy is thus an instrument that implements national policy and strategy. Successful commanders understand the relationship and links between policy and strategy. They also appreciate the distinctions and interrelationships among the levels of war. This appreciation is fundamental to an informed understanding of the decisions and actions at each level. Without it, commanders cannot sequence and synchronize military and nonmilitary actions toward an end state consistent with national strategy and policy.

OPERATIONAL

6-10. The operational level links employing tactical forces to achieving the strategic end state. At the operational level, commanders conduct campaigns and major operations to establish conditions that define that end state. A *campaign* is a series of related major operations aimed at achieving strategic and operational objectives within a given time and space (JP 5-0). A *major operation* is a series of tactical actions (battles, engagements, strikes) conducted by combat forces of a single or several Services, coordinated in time and place, to achieve strategic or operational objectives in an operational area. These actions are conducted simultaneously or sequentially in accordance with a common plan and are controlled by a single commander (JP 3-0). Major operations are not solely the purview of combat forces. They are typically conducted with the other instruments of national power. Major operations often bring together the capabilities of other agencies, nations, and organizations. For noncombat operations, *major operation* refers to the relative size and scope of a military operation (JP 3-0).

6-11. Operational art determines when, where, and for what purpose commanders employ major forces. Operational commanders position and maneuver forces to shape conditions in their area of operations for their decisive operation. Commanders exploit tactical victories to gain strategic advantage or reverse the strategic effects of tactical losses.

6-12. Actions at the operational level usually involve broader dimensions of time and space than tactical actions do. Operational commanders need to understand the complexities of the operational environment, look beyond the immediate situation, and consider the consequences of their approach and subordinates' actions. Operational commanders seek to create the most favorable conditions possible for subordinate commanders by shaping future events.

6-13. Experienced operational commanders understand tactical realities and can create conditions that favor tactical success. Likewise, good tactical commanders understand the operational and strategic context within which they execute their assigned tasks. This understanding helps them seize opportunities (both foreseen and unforeseen) that contribute to establishing the end state or defeating enemy initiatives that threaten its achievement. Operational commanders require experience at both the operational and tactical levels. This experience gives them the knowledge and intuition needed to understand how tactical and operational possibilities interrelate.

TACTICAL

6-14. *Tactics* is the employment and ordered arrangement of forces in relation to each other (CJCSI 5120.02A). Through tactics, commanders use combat power to accomplish missions. The tactical-level commander uses combat power in battles, engagements, and small-unit actions. **A *battle* consists of a set of related engagements that lasts longer and involves larger forces than an engagement**. Battles can affect the course of a campaign or major operation. An *engagement* is a tactical conflict, usually between opposing, lower echelons maneuver forces (JP 1-02). Engagements are typically conducted at brigade level and below. They are usually short, executed in terms of minutes, hours, or days.

6-15. Operational-level headquarters determine objectives and provide resources for tactical operations. For any tactical-level operation, the surest measure of success is its contribution to achieving the end state conditions. Commanders avoid battles and engagements that do not contribute to achieving the operational end state conditions.

APPLYING OPERATIONAL ART

6-16. Commanders use operational art to envision how to create conditions that define the national strategic end state. Actions and interactions across the levels of war influence these conditions. These conditions are fundamentally dynamic and linked together by the human dimension, the most unpredictable and uncertain element of conflict. The operational environment is complex, adaptive, and interactive. Through operational art, commanders apply a comprehensive understanding of it to determine the most effective and efficient methods to influence conditions in various locations across multiple echelons. (See figure 6-2.)

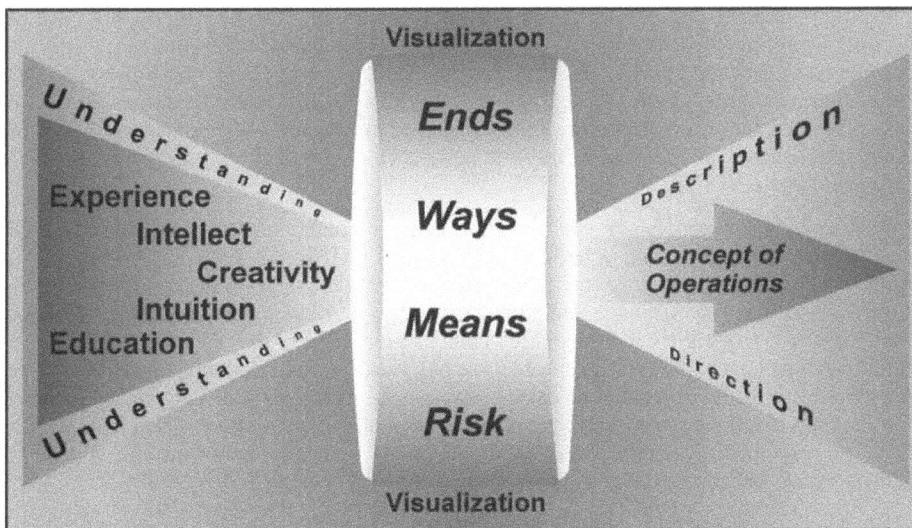

Figure 6-2. Operational art

6-17. Operational art spans a continuum—from comprehensive strategic direction to concrete tactical actions. Bridging this continuum requires creative vision coupled with broad experience and knowledge. Operational art provides a means for commanders to derive the essence of an operation. Without it, tactical actions devolve into a series of disconnected engagements, with relative attrition the only measure of success. Through operational art, commanders translate their concept of operations into an operational design and ultimately into tactical tasks. They do this by integrating ends, ways, and means and by envisioning dynamic combinations of the elements of full spectrum operations across the levels of war. They then apply operational art to array forces and maneuver them to achieve the desired end state.

6-18. Every operation begins with a commander's intent that guides its conduct. In almost all cases, an operational commander's intent and concept of operations envision all the instruments of national power working toward a common end state. Through operational art, commanders frame their concept by answering several fundamental questions:

- What is the force trying to accomplish (ends)?
- What conditions, when established, constitute the desired end state (ends)?
- How will the force achieve the end state (ways)?
- What sequence of actions is most likely to attain these conditions (ways)?

- What resources are required, and how can they be applied to accomplish that sequence of actions (means)?
- What risks are associated with that sequence of actions, and how can they be mitigated (risk)?

6-19. Operational art is generally the purview of joint force commanders. Joint force commanders visualize, describe, and direct all aspects of campaigns. This is particularly true when translating broad national interests and policy into a clearly defined, decisive, and attainable end state and its supporting military conditions. These conditions describe how joint force commanders visualize the state of their operational area upon achieving that end state. Based on a comprehensive analysis of the operational environment, joint force commanders determine the centers of gravity around which to frame the campaign.

6-20. Land component commanders are not directly responsible for defining the military end state. Nonetheless, their participation in the initial stages of campaign design is vital. When designing major operations within a campaign, land component commanders formulate activities at the operational level of war. These land operations heavily influence campaign design because of their complexity, duration, risk, and importance. Typically, military forces directly, routinely, and persistently interact with the populace only in the land domain. Therefore, the land domain is where joint force commanders integrate most capabilities of the other instruments of national power. By introducing the land perspective into the collaborative effort, senior land commanders provide joint force commanders with a broader understanding of landpower's capabilities and limitations. In turn, land force commanders visualize and describe their design of major land operations with a clear understanding of the joint design. This collaboration forms the basis for conducting interdependent joint operations.

6-21. When applying operational art, collaboration informs situational understanding. This collaboration involves an open, continuous dialog between commanders that spans the levels of war and echelons of command. This dialog is essential to reducing the tension inherent to command and control across the levels of war. It is vital in establishing a common perspective on the problem and a shared understanding of the operational environment's conditions. This collaboration ensures that the strategic end state remains linked to operational capabilities. It establishes not only what is possible but also what is practical at the operational level. Effective collaboration enables assessment, fosters critical analysis, and anticipates adaptation. It assumes that the strategic end state is not fixed. Collaboration allows operational commanders to recognize and react to changes in the situation. Operational commanders can then adjust operations so tactical actions remain linked to conditions in the operational environment.

6-22. Practicing operational art requires a broad understanding of the operational environment at all levels. It also requires practical creativity and the ability to visualize changes in the operational environment. Operational commanders need to project their visualization beyond the realm of physical combat. They must anticipate the operational environment's evolving military and nonmilitary conditions. Operational art encompasses visualizing the synchronized arrangement and employment of military forces and capabilities to achieve the strategic or operational end state. This creative process requires the ability to discern the conditions required for victory before committing forces to action.

6-23. Balancing the factors of time, space, and force against a set of conditions is more difficult for operational commanders than for tactical commanders. Those factors are usually fixed at the tactical level. At the operational level, the situation is more complex; nonmilitary aspects of the situation dominate and present intangible elements increasingly difficult to quantify. The greater the strategic end state's scope, the more uncertain the situation facing operational commanders.

6-24. Conflict is fundamentally a human endeavor. It is characterized by violence, uncertainty, chance, and friction. Land operations are inherently tied to the human dimension; they cannot be reduced to a simple formula or checklist. Operational art helps commanders integrate diverse capabilities, including those related to the human dimension. It also helps synchronize military actions with actions of other instruments of national power. Operational art provides the conceptual framework for ordering thought when designing operations. It is the creative engine that drives commanders' ability to seize, retain, and exploit initiative and achieve decisive results.

THE ELEMENTS OF OPERATIONAL DESIGN

6-25. *Operational design* is the conception and construction of the framework that underpins a campaign or major operation plan and its subsequent execution (JP 3-0). Through operational art, commanders and staffs develop a broad concept for applying the military instrument, including landpower, and translate it into a coherent, feasible design for employing joint forces. This operational design provides a framework that relates tactical tasks to the strategic end state. It provides a unifying purpose and focus to all operations. (See figure 6-3.)

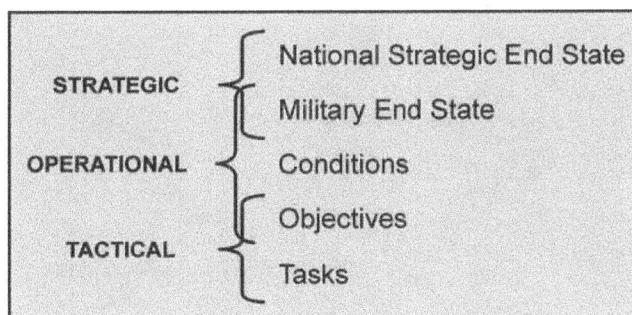

Figure 6-3. Relating tactical tasks to the strategic end state

6-26. In applying operational art, commanders consider the elements of operational design. These are tools to help clarify and refine their concept of operations by providing a framework to describe its execution. They help commanders understand, visualize, and describe complex combinations of combat power and help them formulate their intent and guidance. The elements of operational design may be used selectively in any operation. However, their application is broadest in the context of campaigns and major operations.

6-27. The elements of operational design are essential to identifying tasks and objectives that tie tactical missions to achieving the strategic end state. They help refine and focus the concept of operations that forms the basis for developing a detailed plan or order. During execution, commanders and staffs consider the design elements as they assess the situation. They adjust current and future operations and plans as the operation unfolds.

6-28. Commanders and staffs gauge how the elements of operational design relate to the mission variables. The applicability of individual elements varies with echelon. Generally, all apply at the strategic and operational levels. Some have no tactical relevance whatsoever. For example, land component commanders translate operational reach and the culminating point into a broad design for operational maneuver; decisive points become objectives along lines of operations and lines of effort. In contrast, leaders at lower tactical echelons may only consider objectives. Ultimately, commanders at each echelon determine which elements are relevant, based on the mission and conditions.

6-29. Operational art requires three continuous, cyclic activities. (See figure 6-4.) These activities define military and nonmilitary actions across the spectrum of conflict:
- Framing (and reframing) the problem.
- Formulating the design.
- Refining the design.

Individually, these activities further refine the nature of the operation, determine force structure, focus operations, and prioritize resources. Together, they help operational commanders translate broad political guidance into tangible tactical actions. The elements of operational design help commanders develop planning guidance and communicate their intent. These actions are the first steps in setting the conditions for operational and tactical success.

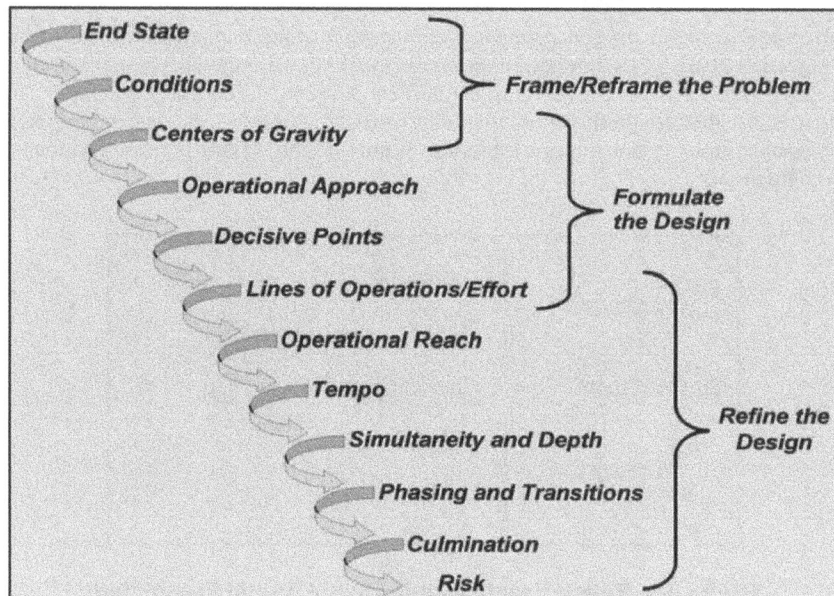

Figure 6-4. Linking the elements of operational design

FRAMING THE PROBLEM

6-30. Framing (and reframing) the problem encompasses receiving and clarifying the strategic end state and conditions. It also includes determining the appropriate operational theme, operational approach, and defeat or stability mechanisms. Framing requires understanding and viewing the operational environment from a systemic perspective and identifying and analyzing centers of gravity. Collaboration informs problem framing.

End State and Conditions

6-31. At the national strategic level, the end state is represented by the broadly expressed conditions that the President wants to exist when a campaign or major operation ends. The end state is thus an image of the operational environment consistent with national interests and policy. The strategic end state is achieved through the integrated, collective activities of all instruments of national power, not by any single instrument applied in isolation.

6-32. Exercising battle command at all levels requires understanding the operational environment's conditions. Clearly describing the end state requires understanding the operational environment and operational theme, and assessing the environment's friendly, enemy, adversary, and neutral aspects. Commanders include it in their planning guidance and commander's intent. A clearly defined end state promotes unity of effort, facilitates integration and synchronization, and helps mitigate risk.

6-33. Based on the strategic end state, commanders determine the conditions that compose the supporting military end state. Army operations typically focus on attaining the military end state. However, Army operations also contribute to establishing nonmilitary conditions. Sometimes that is their focus. Operational-level commanders explicitly describe the end state and its defining conditions for every operation. Otherwise, the necessary integration between tactical tasks and operational conditions does not occur. Missions become vague and operations lose focus. Every operation should be directed toward a clearly defined, decisive, and attainable end state.

6-34. The end state may evolve as a campaign progresses. Strategic and operational guidance may be refined, the operational environment's conditions might change, and situational understanding may increase. Therefore, all commanders continuously monitor operations and evaluate their progress. Operational commanders assess progress against measures of effectiveness and the end state conditions. These conditions form the basis for decisions that ensure operations progress consistently toward the desired end state.

Centers of Gravity

6-35. A *center of gravity* is the source of power that provides moral or physical strength, freedom of action, or will to act (JP 3-0). This definition states in modern terms the classic description offered by Clausewitz: "the hub of all power and movement, on which everything depends."[4] The loss of a center of gravity ultimately results in defeat. The center of gravity is a vital analytical tool for designing campaigns and major operations. It provides a focal point for them, identifying sources of strength and weakness at the strategic and operational levels of war.

6-36. Modern understanding of the center of gravity has evolved beyond the term's preindustrial definition. Centers of gravity are now part of a more complex perspective of the operational environment. Today they are not limited to military forces and can be either physical or moral. Physical centers of gravity, such as a capital city or military force, are typically easier to identify, assess, and target. They can often be influenced solely by military means. In contrast, moral centers of gravity are intangible and complex. They are dynamic and related to human factors. Examples include a charismatic leader, powerful ruling elite, religious tradition, tribal influence, or strong-willed populace. Military means alone are usually ineffective when targeting moral centers of gravity. Eliminating them requires the collective, integrated efforts of all instruments of national power.

6-37. Typically, a single enemy center of gravity, and a single friendly center of gravity exist at the strategic and operational levels of war for a given campaign or major operation. However, in some campaigns none exist below the strategic level; in others, there may be more than a single friendly or enemy center of gravity. Centers of gravity are not relevant at the tactical level; the tactical equivalent is the objective. At the strategic level, the center of gravity may be vulnerable to an operational-level approach; at the operational level, the center of gravity may be vulnerable to tactical actions. The enemy may shift a center of gravity to protect and sustain a source of power. Similarly, changes in the operational environment may cause centers of gravity to shift. Therefore, analysis of friendly and enemy centers of gravity begins during planning and continues throughout a campaign or major operation.

6-38. Center of gravity analysis is thorough and detailed. Faulty conclusions drawn from hasty or abbreviated analyses can adversely affect operations, waste critical resources, and incur undue risk. A thorough understanding of the operational environment facilitates identifying and targeting enemy centers of gravity. This understanding encompasses how enemies and adversaries organize, fight, and make decisions. It also includes their physical and moral strengths and weaknesses. In addition, commanders should understand how military forces interact with the other operational variables. This understanding helps planners identify centers of gravity, their associated decisive points, and the best operational approach for isolating, neutralizing, and defeating them.

6-39. Thorough and detailed analyses help commanders and staffs understand the operational environment's complex nature. From this understanding, they determine the actions necessary to achieve the desired end state. As sources of power, centers of gravity are inherently complex. Commanders and staffs examine them systemically to determine the relationships between the conditions and resources that enable each one to function. Understanding the dynamic, complex nature of a center of gravity is the key to exposing its vulnerabilities. Commanders and staffs identify vulnerabilities that expose enemy centers of gravity to paralysis, shock, and collapse. Friendly forces can then attack those vulnerabilities and set conditions for success.

---

[4] © 1984. Reproduced with permission of Princeton University Press.

FORMULATING THE DESIGN

6-40. Formulating the design builds on the framework developed previously. It continues the analysis of centers of gravity and explores the nature of the enemy or adversary by—

- Determining the appropriate operational approach.
- Identifying decisive points.
- Devising lines of operations and effort.

Determining the operational approach includes identifying the defeat or stability mechanisms that best accomplish the mission. Decisive points that offer the greatest leverage against centers of gravity are then selected. Finally, commanders determine the combination of lines of operations and effort that best translate the concept of operations into specific tactical tasks.

Operational Approach

6-41. **The *operational approach* is the manner in which a commander contends with a center of gravity**. There are two operational approaches: direct and indirect. **The *direct approach* attacks the enemy's center of gravity or principal strength by applying combat power directly against it**. However, centers of gravity are generally well protected and not vulnerable to a direct approach. Thus, commanders usually choose an indirect approach. **The *indirect approach* attacks the enemy's center of gravity by applying combat power against a series of decisive points while avoiding enemy strength**. Both approaches use unique combinations of defeat or stability mechanisms, depending on the mission. Whether direct or indirect, an effective operational approach achieves decisive results through combinations of defeat and stability mechanisms. As commanders and staffs frame the problem, they determine the appropriate combination of defeat or stability mechanisms to solve it. This begins the process that ends with the design for an operation that achieves the desired end state.

Defeat Mechanisms

6-42. **A *defeat mechanism* is the method through which friendly forces accomplish their mission against enemy opposition**. A defeat mechanism is described in terms of the physical or psychological effects it produces. Defeat mechanisms are not tactical missions; rather, they describe broad operational and tactical effects. Commanders must translate these effects into tactical tasks. Operational art formulates the most effective, efficient way to defeat enemy aims. Physical defeat deprives enemy forces of the ability to achieve those aims; psychological defeat deprives them of the will to do so. Army forces are most successful when applying focused combinations of defeat mechanisms. This produces complementary and reinforcing effects not attainable with a single mechanism. Used individually, a defeat mechanism achieves results proportional to the effort expended. Used in combination, the effects are likely to be both synergistic and lasting. Army forces at all echelons use combinations of four defeat mechanisms:

- Destroy.
- Dislocate.
- Disintegrate.
- Isolate.

6-43. ***Destroy* means to apply lethal combat power on an enemy capability so that it can no longer perform any function and cannot be restored to a usable condition without being entirely rebuilt**. The most effective way to destroy enemy capabilities is with a single, decisive attack. When the necessary combat power cannot be massed simultaneously, commanders apply it sequentially. This approach is called attrition. It defeats the enemy by maintaining the highest possible rate of destruction over time.

6-44. Destruction may not force the enemy to surrender; well-disciplined forces and those able to reconstitute can often endure heavy losses without giving up. Defeat cannot be accurately measured solely in terms of destruction. This is particularly true when criteria focus on narrow metrics, such as casualties, equipment destroyed, or perceived enemy strength. Destruction is especially difficult to assess if friendly forces apply force indiscriminately. The effects of destruction are often transitory unless combined with isolation and dislocation.

6-45. ***Dislocate* means to employ forces to obtain significant positional advantage, rendering the enemy's dispositions less valuable, perhaps even irrelevant**. It aims to make the enemy expose forces by reacting to the dislocating action. Dislocation requires enemy commanders to make a choice: accept neutralization of part of their force or risk its destruction while repositioning. Turning movements and envelopments produce dislocation. When combined with destruction, dislocation can contribute to rapid success.

6-46. ***Disintegrate* means to disrupt the enemy's command and control system, degrading the ability to conduct operations while leading to a rapid collapse of the enemy's capabilities or will to fight**. It exploits the effects of dislocation and destruction to shatter the enemy's coherence. Typically, disintegration follows the loss of capabilities that enemy commanders use to develop and maintain situational understanding, coupled with destruction and dislocation. Simultaneous operations produce the strongest disintegrative effects. Disintegration is difficult to achieve; however, prolonged isolation, destruction, and dislocation can produce it.

6-47. ***Isolate* means to deny an enemy or adversary access to capabilities that enable the exercise of coercion, influence, potential advantage, and freedom of action**. Isolation limits the enemy's ability to conduct operations effectively by marginalizing one or more of these capabilities. It exposes the enemy to continued degradation through the massed effects of the other defeat mechanisms. There are two types of isolation:

- Physical isolation, which is difficult to achieve, but easier to assess. An isolated enemy loses freedom of movement and access to support.
- Psychological isolation, which, while difficult to assess, is a vital enabler of disintegration. The most important indicators include the breakdown of enemy morale and the alienation of a population from the enemy.

6-48. Isolation alone rarely defeats an enemy. However, it complements and reinforces other defeat mechanisms' effects. Offensive operations often focus on destroying personnel and equipment. They may use maneuver to dislocate forces. However, these effects multiply when combined with isolating the enemy from sources of physical and moral support.

Stability Mechanisms

6-49. Commanders use stability mechanisms to visualize how to employ the stability element of full spectrum operations. A ***stability mechanism* is the primary method through which friendly forces affect civilians in order to attain conditions that support establishing a lasting, stable peace**. As with defeat mechanisms, combinations of stability mechanisms produce complementary and reinforcing effects that accomplish the mission more effectively and efficiently than single mechanisms do alone. The four stability mechanisms are—

- Compel.
- Control.
- Influence.
- Support.

6-50. ***Compel* means to use, or threaten to use, lethal force to establish control and dominance, effect behavioral change, or enforce compliance with mandates, agreements, or civil authority**. The appropriate and discriminate use of lethal force reinforces efforts to stabilize a situation, gain consent, or assure compliance. Conversely, misusing force can adversely affect an operation's legitimacy. Legitimacy is essential to producing effective compliance. Compliance depends on how the local populace and others perceive the force's ability to exercise lethal force to accomplish the mission.

6-51. In this context, ***control* means to impose civil order**. It includes securing borders, routes, sensitive sites, population centers, and individuals. It also involves physically occupying key terrain and facilities. Control includes activities related to disarmament, demobilization, and reintegration, as well as security sector reform. (Paragraphs 3-89 through 3-90 discuss the security sector.)

6-52. ***Influence* means to alter the opinions and attitudes of a civilian population through information engagement, presence, and conduct**. It aims to change behaviors through nonlethal means. Influence is as much a product of public perception as a measure of operational success. It reflects the ability of friendly forces to operate within the cultural and societal norms of the local populace while accomplishing the mission. Influence requires legitimacy. Developing legitimacy requires time, patience, and coordinated, cooperative efforts across the operational area.

6-53. *Support* **means to establish, reinforce, or set the conditions necessary for the other instruments of national power to function effectively.** It requires coordination and cooperation with civilian agencies as they assess the immediate needs of failed or failing states and plan, prepare for, or execute responses to them. In extreme circumstances, support may require committing considerable resources for a protracted period. This commitment may involve establishing or reestablishing the institutions required for normal life. These typically include a legitimate civil authority, market economy, and criminal justice system supported by government institutions for health, education, and civil service.

Using Defeat and Stability Mechanisms

6-54. Defeat and stability mechanisms complement center of gravity analysis. This analysis helps to frame an operational-level problem; defeat and stability mechanisms suggest means to solve it. The analysis reveals the intrinsic vulnerabilities of a given center of gravity; defeat mechanisms describe ways to isolate, weaken, or destroy it. For example, a decisive point may be temporarily neutralized by dislocating it. The enemy may commit significant combat power to regain that capability, presenting an opportunity to destroy committed enemy forces. By combining dislocation and destruction, the commander can effectively eliminate the capability. Thus, the effect on the center of gravity is permanent, and friendly forces retain freedom of action and initiative.

6-55. The operational approach reflects the commander's visualization for applying combinations of defeat and stability mechanisms. An effective operational approach, direct or indirect, focuses operations toward establishing the end state.

Decisive Points

6-56. A *decisive point* is a geographic place, specific key event, critical factor, or function that, when acted upon, allows commanders to gain a marked advantage over an adversary or contribute materially to achieving success (JP 3-0). Decisive points are not centers of gravity; they are keys to attacking or protecting them. Decisive points apply at both the operational and tactical levels. At the operational level, they typically provide direct leverage against a center of gravity. At the tactical level, they are directly tied to mission accomplishment.

6-57. Some decisive points are geographic. Examples include port facilities, distribution networks and nodes, and bases of operations. Specific events and elements of an enemy force may also be decisive points. Examples of such events include commitment of the enemy operational reserve and reopening a major oil refinery. A common characteristic of decisive points is their major importance to a center of gravity. A decisive point's importance requires the enemy to commit significant resources to defend it. The loss of a decisive point weakens a center of gravity and may expose more decisive points.

6-58. Decisive points have a different character during operations dominated by stability or civil support. These decisive points may be less tangible and more closely associated with important events and conditions. For example, during operations after Hurricane Andrew in 1992, reopening schools was a decisive point. Other examples include—

- Repairing a vital water treatment facility.
- Establishing a training academy for national security forces.
- Securing an election.
- Quantifiably reducing crime.

None of these examples is purely physical. Nonetheless, any may be vital to establishing conditions for transitioning to civil authority. In an operation dominated by stability or civil support, this transition is typically an end state condition.

6-59. Commanders identify the decisive points that offer the greatest leverage against centers of gravity. Decisive points that enable commanders to seize, retain, or exploit the initiative are crucial. Controlling them is essential to mission accomplishment. Enemy control of a decisive point may exhaust friendly momentum, force early culmination, or allow an enemy counterattack. Decisive points shape the design of operations. They help commanders select clearly decisive, attainable objectives that directly contribute to establishing the end state.

Lines of Operations and Lines of Effort

6-60. In an operational design, lines of operations and lines of effort bridge the broad concept of operations across to discreet tactical tasks. They link tactical and operational objectives to the end state. Continuous assessment gives commanders the information required to revise and adjust lines of operations and effort. Subordinate commanders reallocate resources accordingly.

6-61. Commanders may describe an operation along lines of operations, lines of effort, or a combination of both. Irregular warfare, for example, typically requires a deliberate approach using lines of operations complemented with lines of effort; the combination of them may change based on the conditions within the operational area. An operational design using both lines of operations and lines of effort reflects the characteristics and advantages of each. With this approach, commanders synchronize and sequence actions, deliberately creating complementary and reinforcing effects. The lines then converge on the well-defined, commonly understood end state outlined in the commander's intent.

Lines of Operations

6-62. A *line of operations* is a line that defines the directional orientation of a force in time and space in relation to the enemy and links the force with its base of operations and objectives. (See figure 6-5.) Lines of operations connect a series of decisive points that lead to control of a geographic or force-oriented objective. Operations designed using lines of operations generally consist of a series of actions executed according to a well-defined sequence. Major combat operations are typically designed using lines of operations. These lines tie offensive and defensive tasks to the geographic and positional references in the operational area. Commanders synchronize activities along complementary lines of operations to achieve the end state. Lines of operations may be either interior or exterior.

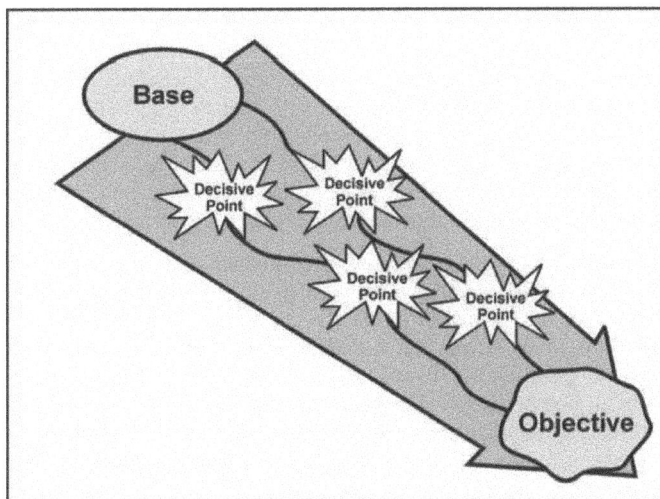

Figure 6-5. Example of a line of operations

6-63. A force operates on *interior lines* when its operations diverge from a central point. Interior lines usually represent central position, where a friendly force can reinforce or concentrate its elements faster than the enemy force can reposition. With interior lines, friendly forces are closer to separate enemy forces than the enemy forces are to one another. Interior lines allow an isolated force to mass combat power against a specific portion of an enemy force by shifting capabilities more rapidly than the enemy can react.

6-64. **A force operates on *exterior lines* when its operations converge on the enemy**. Operations on exterior lines offer opportunities to encircle and annihilate an enemy force. However, these operations typically require a force stronger or more mobile than the enemy.

6-65. The relevance of interior and exterior lines depends on the time and space relationship between the opposing forces. Although an enemy force may have interior lines with respect to the friendly force, this advantage disappears if the friendly force is more agile and operates at a higher tempo. (Paragraph 6-80 defines tempo.) Conversely, if a smaller friendly force maneuvers to a position between larger but less agile enemy forces, the friendly force may be able to defeat them in detail before they can react effectively.

Lines of Effort

6-66. **A *line of effort* links multiple tasks and missions using the logic of purpose—cause and effect— to focus efforts toward establishing operational and strategic conditions**. Lines of effort are essential to operational design when positional references to an enemy or adversary have little relevance. In operations involving many nonmilitary factors, lines of effort may be the only way to link tasks, effects, conditions, and the desired end state. Lines of effort are often essential to helping commanders visualize how military capabilities can support the other instruments of national power. They are a particularly valuable tool when used to achieve unity of effort in operations involving multinational forces and civilian organizations, where unity of command is elusive, if not impractical.

6-67. Commanders use lines of effort to describe how they envision their operations creating the more intangible end state conditions. These lines of effort show how individual actions relate to each other and to achieving the end state. Ideally, lines of effort combine the complementary, long-term effects of stability or civil support tasks with the cyclic, short-term events typical of offensive or defensive tasks.

6-68. Commanders at all levels may use lines of effort to develop missions and tasks and to allocate resources. Commanders may designate one line of effort as the decisive operation and others as shaping operations. Commanders synchronize and sequence related actions along multiple lines of effort. Seeing these relationships helps commanders assess progress toward achieving the end state as forces perform tasks and accomplish missions.

6-69. Commanders typically visualize stability and civil support operations along lines of effort. For stability operations, commanders may consider linking primary stability tasks to their corresponding Department of State post-conflict technical sectors. (See paragraphs 3-88 through 3-95.) These stability tasks link military actions with the broader interagency effort across the levels of war. Figure 6-6, page 6-14, provides an example. A full array of lines of effort might include offensive and defensive lines, as well as a line for information operations. Information operations typically produce effects across multiple lines of effort.

6-70. The five post-conflict technical sectors described in chapter 3 may become lines of effort, as illustrated in figure 6-6. They provide a framework for analyzing an operational environment where stability operations are the major focus. They identify the breadth and depth of relevant civilian agency tasks and emphasize the relationships among them. Using them as lines of effort can help as Army forces in collaborative interagency planning and dialog. Planning and dialog leads to developing lines of effort that synchronize the effects of all instruments of national power. However, the sectors are not themselves lines of effort and are not the default solution for every operation. During a civil support mission, lines of effort normally portray support in response to a disaster or terrorist attack, support to law enforcement, and other support as required.

6-71. As operations progress, commanders may modify the lines of effort after assessing conditions and collaborating with multinational military and civilian partners. Lines of effort typically focus on integrating the effects of military operations with those of other instruments of national power to support the broader effort. Each operation, however, is different. Commanders develop and modify lines of effort to keep operations focused on achieving the end state, even as the situation changes.

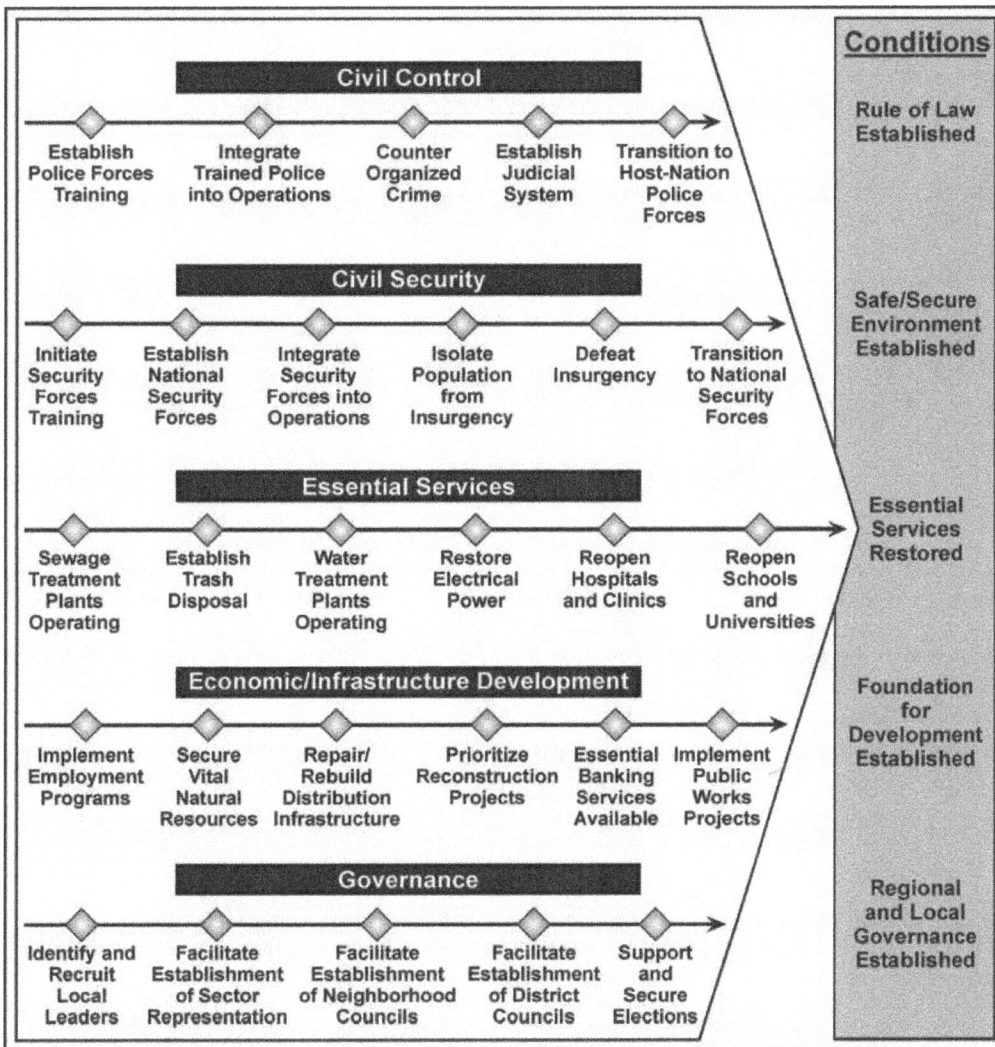

Figure 6-6. Example of lines of effort (stability)

Combining Lines of Operations and Lines of Effort

6-72. Commanders use both lines of operations and lines of effort to connect objectives to a central, unifying purpose. Lines of operations portray the more traditional links between objectives, decisive points, and centers of gravity. However, lines of operations do not project the operational design beyond defeating enemy forces and seizing terrain. Combining lines of operations and lines of effort allows commanders to include nonmilitary activities in their operational design. This combination helps commanders incorporate stability tasks that set the end state conditions into the operation. It allows commanders to consider the less tangible aspects of the operational environment where the other instruments of national power dominate. Commanders can then visualize concurrent and post-conflict stability activities. Making these connections relates the tasks and purposes of the elements of full spectrum operations with joint effects identified in the campaign plan. The resulting operational design effectively combines full spectrum operations throughout the campaign or major operation.

REFINING THE DESIGN

6-73. Refining the design examines how commanders should execute the operation in the operational environment. Commanders determine their force's operational reach and culminating point. They also determine the tempo that will allow them to accomplish the mission without culminating. They consider several aspects of simultaneity and depth. These considerations include the emphasis each element of full spectrum operations will receive and the extent to which the force can operate throughout the operational area. When necessary, commanders phase the operation and plan transitions to maintain momentum. Throughout this activity, commanders consider the risks they are willing to accept.

Operational Reach

> *The third factor, the distance from the sources that must send continual replacements for this steadily weakening army, will increase proportionately with the advance. In this respect a conquering army is like the light of a lamp; as the oil that feeds it sinks and draws away from the focus, the light diminishes until at last it goes out altogether.*
>
> Carl von Clausewitz
> *On War*[5]

6-74. *Operational reach* is the distance and duration across which a unit can successfully employ military capabilities (JP 3-0). It reflects the ability to achieve success through a well-conceived operational approach. Operational reach is a tether; it is a function of protection, sustainment, endurance, and relative combat power. The limit of a unit's operational reach is its culminating point. (See paragraph 6-94.) Operational design extends operational reach while creating the conditions necessary to seize, retain, and exploit the initiative. It balances the natural tension between endurance, momentum, and protection. The following actions can extend operational reach:

- Forward positioning of forces, reserves, bases, and support capabilities along lines of operations.
- Employing weapons systems with extended ranges, such as missiles.
- Phasing an operation to focus limited resources.
- Leveraging supply discipline, contracting, and host-nation support. Maximizing distribution network efficiency.
- Leveraging joint capabilities.

6-75. Endurance is the ability to employ combat power anywhere for protracted periods. It stems from the ability to generate, protect, and sustain a force, regardless of the distance from its base and the austerity of the environment. Endurance involves anticipating requirements and making the most effective, efficient use of available resources. Their endurance gives Army forces their campaign capability. It makes permanent the transitory effects of other capabilities.

6-76. Momentum comes from seizing the initiative and executing high-tempo operations that overwhelm enemy resistance. Commanders control momentum by maintaining focus and pressure. They set a tempo that prevents exhaustion and maintains sustainment. A sustainable tempo extends operational reach. Commanders maintain momentum by anticipating and transitioning rapidly between primary tasks or—when necessary—the elements of full spectrum operations. Sometimes commanders must push the force to its culminating point to take maximum advantage of an opportunity. For example, exploitations and pursuits often involve pushing all available forces to the limit of their endurance to capitalize on momentum and retain the initiative.

6-77. Protection is an important contributor to operational reach. Commanders anticipate how enemy actions might disrupt operations and then determine the protection capabilities required to maintain sufficient reach. Protection closely relates to endurance and momentum. It also contributes to the commander's

---

[5] © 1984. Reproduced with permission of Princeton University Press.

ability to extend operations in time and space. The protection warfighting function helps commanders maintain the deploying force's integrity and combat power.

6-78. **An *operational pause* is a deliberate halt taken to extend operational reach or prevent culmination.** Commanders may execute an operational pause for several reasons. These may include the force being close to culmination, the decisive operation failing, or the strategic end state changing. In designing an operation, commanders carefully balance initiative, momentum, reach, and culmination to avoid unnecessary operational pauses. In protracted operations, however, they may have to execute operational pauses to extend operational reach. Commanders carefully plan, prepare for, and execute these pauses to prevent losing the initiative. During an operational pause, commanders retain the initiative by using shaping operations to keep pressure on enemy forces. These shaping operations confuse the enemy while friendly forces generate combat power for the decisive operation.

6-79. Operational design balances operational reach, operational approach, and operational pauses to ensure Army forces accomplish their missions before culminating. Commanders continually strive to extend operational reach. They assess friendly and enemy force status, anticipate culmination, and plan operational pauses if necessary. Since the time of Sun Tzu, commanders have studied and reflected on the challenge of conducting and sustaining operations over long distances and times. History contains many examples of campaigns hampered by inadequate operational reach. Achieving the desired end state requires forces with the operational reach to establish and maintain the end state conditions.

Tempo

6-80. ***Tempo* is the relative speed and rhythm of military operations over time with respect to the enemy.** It reflects the rate of military action. Controlling tempo helps commanders keep the initiative during combat operations or rapidly establish a sense of normalcy during humanitarian crises. During operations dominated by the offense and defense, commanders normally seek to maintain a higher tempo than the enemy; rapid tempo can overwhelm an enemy's ability to counter friendly actions. During operations dominated by stability and civil support, commanders act quickly to control events. By acting faster than the situation deteriorates, commanders can change the dynamics of a crisis and restore stability. The capability to act quickly enhances flexibility and adaptability across the spectrum of conflict.

6-81. Commanders control tempo through deliberate operational design. First, they formulate operations that stress the complementary and reinforcing effects of simultaneous and sequential operations. They synchronize those operations in time and space to degrade enemy capabilities throughout the operational area. Second, commanders avoid unnecessary engagements. This practice includes bypassing resistance that appears at times and places commanders do not consider decisive. Third, through mission command they give subordinates maximum latitude to exercise initiative and act independently. Controlling tempo requires both audacity and patience: audacity initiates the actions needed to develop a situation; patience allows a situation to develop until the force can strike at the decisive time and place. Ultimately, the goal is maintaining a tempo appropriate to retaining the initiative and achieving the end state.

6-82. Army forces expend more energy and resources when operating at a high tempo. Commanders assess the force's capacity to operate at a high tempo based on its performance and available resources. An effective operational design varies tempo throughout an operation to increase endurance while maintaining appropriate speed and momentum. There is more to tempo than speed. While speed can be important, commanders balance speed with endurance.

Simultaneity and Depth

6-83. Simultaneity and depth extend operations in time and space. Simultaneity has two components. Both depend on depth to attain lasting effects and maximum synergy. Simultaneous combinations of offensive, defensive, and stability tasks overwhelm enemy forces and their will to resist while setting the conditions for a lasting, stable peace. Simultaneous actions across the depth of the operational area place more demands on enemy forces than they can effectively respond to them. Operations combining depth and simultaneity achieve a synergy that paralyzes enemy forces. This prevents them from reacting appropriately, inducing their early culmination. Similarly, stability or civil support tasks—executed in depth with simultaneous defensive and offensive tasks when necessary—establish control of the situation throughout the operational area.

6-84. Simultaneity also refers to the concurrent effects operations produce at the tactical, operational, and strategic levels. Tactical commanders fight the battles and engagements that achieve objectives in accordance with the operational commander's intent. Operational commanders set the conditions for tactical success in battles within a campaign or major operation. These victories, in turn, create the conditions that define the end state. Because of the complex interaction among the levels of war, commanders cannot be concerned only with events at their respective echelon. Success requires them to understand how their actions affect the operations of commanders at all other echelons.

6-85. *Depth* **is the extension of operations in time, space, and resources**. Operations in depth can disrupt the enemy's decision cycle. These operations contribute to protecting the force by destroying enemy capabilities before the enemy can use them. Commanders balance their forces' tempo to produce simultaneous results throughout their operational area. To achieve simultaneity, commanders establish a higher tempo to target enemy capabilities located at the limit of a force's operational reach.

6-86. Simultaneity and depth are inherent in full spectrum operations. Army forces execute simultaneous operations across vast areas. They force the enemy to react to numerous friendly actions—potential and actual—throughout the operational area. Army forces use combined arms, advanced information systems, and joint capabilities to increase the depth of their operations. The complementary effects produced by executing simultaneous operations in depth overwhelm enemy forces, forcing them to respond piecemeal or not at all.

6-87. Commanders extend the depth of operations through joint integration. When determining an operation's depth, commanders consider their own capabilities, as well as joint capabilities and limitations. They use these capabilities to ensure actions executed at operational depth receive robust and uninterrupted support. Commanders sequence and synchronize operations in time and space to achieve simultaneous effects throughout the operational area.

Phasing and Transitions

6-88. **A** *phase* **is a planning and execution tool used to divide an operation in duration or activity. A change in phase usually involves a change of mission, task organization, or rules of engagement. Phasing helps in planning and controlling and may be indicated by time, distance, terrain, or an event**. The ability of Army forces to extend operations in time and space, coupled with a desire to dictate tempo, often presents commanders with more objectives and decisive points than the force can engage simultaneously. At both the operational and tactical levels, this situation may require commanders and staffs to consider sequencing operations. Sequencing involves integrating capabilities and synchronizing actions within the operational design. This is accomplished through phasing.

6-89. Phasing is key to arranging complex operations. It describes how the commander envisions the overall operation unfolding. It is the logical expression of the commander's visualization in time. Within a phase, a large portion of the force executes similar or mutually supporting activities. Achieving a specified condition or set of conditions typically marks the end of a phase.

6-90. Simultaneity, depth, and tempo are vital to full spectrum operations. However, they cannot always be attained to the degree desired. In such cases, commanders limit the number of objectives and decisive points engaged simultaneously. They deliberately sequence certain actions to maintain tempo while focusing combat power at the decisive point in time and space. Commanders combine simultaneous and sequential operations to achieve the end state conditions.

6-91. Phasing can extend operational reach. Only when the force lacks the capacity to accomplish the mission in a single action do commanders phase the operation. Each phase should strive to—

- Focus effort.
- Concentrate combat power in time and space at a decisive point.
- Deliberately and logically achieve its objectives.

6-92. Transitions mark a change of focus between phases or between the ongoing operation and execution of a branch or sequel. Shifting priorities between the elements of full spectrum operations—such as from offense to stability—also involves a transition. Transitions require planning and preparation well before their execution to maintain the momentum and tempo of operations. The force is vulnerable during transitions, and commanders establish clear conditions for their execution. Transitions may create unexpected opportunities; they may also make forces vulnerable to enemy threats.

6-93. An unexpected change in conditions may require commanders to direct an abrupt transition between phases. In such cases, the overall composition of the force remains unchanged despite sudden changes in mission, task organization, and rules of engagement. Typically, task organization evolves to meet changing conditions; however, transition planning must also account for changes in mission. Commanders attuned to sudden changes can better adapt their forces to dynamic conditions. They continuously assess the situation and task-organize and cycle their forces to retain the initiative. They strive to achieve changes in emphasis without incurring an operational pause.

Culmination

6-94. **The *culminating point* is that point in time and space at which a force no longer possesses the capability to continue its current form of operations**. Culmination represents a decisive shift in relative combat power. It is relevant to both attackers and defenders at each level of war. In the offense, the culminating point occurs when the force cannot continue the attack and must assume a defensive posture or execute an operational pause. In the defense, it occurs when the force can no longer defend itself and must withdraw or risk destruction.

6-95. With stability, the culminating point is more difficult to identify. Three conditions can result in culmination:

● Being too dispersed to adequately control the situation.

● Being unable to provide the necessary security.

● Lacking required resources.

6-96. During civil support, culmination is unlikely. However, culmination may occur if forces must respond to more catastrophic events than they can manage simultaneously. That situation results in culmination due to exhaustion.

6-97. Strategic culmination occurs when a nation or other opponent no longer has the resources to continue the conflict. Strategic culmination occurs only once during a campaign. It marks when one side can no longer maintain a favorable ratio of military resources and the likelihood for a successful outcome rapidly diminishes. Attackers are compelled to assume the defense or risk defeat. Defenders are forced to sue for peace or face destruction and possibly occupation. Strategic culmination may result from—

● Erosion of domestic national or political will.

● Decline of popular support.

● Lack of focus or other resources.

● Excessive casualties.

6-98. At the operational level, factors influencing culmination are complex. Culminating points at the operational level are linked to time, space, and existing combat power. They directly correspond to the ability to apply available combat power across a set distance within a given period. Operational-level forces are also vulnerable to tactical culmination. The interdependent nature of these factors creates uncertainty. This makes it difficult to anticipate operational and strategic culminating points. The loss of critical capabilities or an unanticipated tactical success can quickly shift the advantage between opposing forces.

6-99. At the tactical level, culmination may be a planned event. In such cases, the concept of operations predicts which part of the force will culminate, and the task organization includes additional forces to assume the mission. Culmination is expected and measures are in place to mitigate it. Tactical culminating points are easier to identify than operational culminating points because they generally occur as a result of a battle or engagement. Tactical culmination is typically caused by direct combat actions or higher echelon resourcing decisions. It relates to the force's ability to generate and apply combat power and is not a lasting condition. Tactical units may be reinforced or reconstituted to continue operations.

Risk

In all great actions there is risk.

Arthur Wellesley, later 1st Duke of Wellington
letter to the Governor General of India, September 1803

6-100. Risk, uncertainty, and chance are inherent in all military operations. When commanders accept risk, they create opportunities to seize, retain, and exploit the initiative and achieve decisive results. Risk is a potent catalyst that fuels opportunity. The willingness to incur operational risk is often the key to exposing enemy weaknesses that the enemy considers beyond friendly reach. Understanding risk requires calculated assessments coupled with boldness and imagination. Successful commanders assess risk continuously throughout operations and mitigate it through creative operational design.

6-101. It is reckless to commit forces without adequate planning and preparation. It is equally rash to delay action while waiting for perfect intelligence and synchronization. Reasonably estimating and intentionally accepting risk is fundamental to conducting operations. It is essential to successful battle command. Successfully applying military force requires commanders who assess the risks, analyze and minimize the hazards, and execute a plan that accounts for those hazards. Experienced commanders balance audacity and imagination with risk and uncertainty to strike at a time and place and in a manner wholly unexpected by enemy forces. This is the essence of surprise. It results from carefully considering and accepting risk. (FMs 3-90 and 6-0 discuss tactical risk.)

6-102. Operational art balances risk and opportunity to create and maintain the conditions necessary to seize, retain, and exploit the initiative and achieve decisive results. During execution, opportunity is fleeting. The surest means to create opportunity is to accept risk while minimizing hazards to friendly forces. A good operational design considers risk and uncertainty equally with friction and chance. The final plans and orders then provide the flexibility commanders need to take advantage of opportunity in complex, dynamic environments.

SUMMARY

6-103. Operational art remains the creative aspect of operations. While the character of conflict changes with time, the violent and chaotic nature of warfare does not. The essence of military art remains timeless. Operational art—the creative expression of informed vision to integrate ends, ways, and means across the levels of war—is fundamental to the Army's ability to seize, retain, and exploit the initiative while concurrently creating and preserving the conditions necessary to restore stability.

This page intentionally left blank.

Chapter 7

Information Superiority

*Be first with the truth. Since Soldier actions speak louder than what [public affairs offi-
cers] say, we must be mindful of the impact our daily interactions with Iraqis have on
global audiences via the news media. Commanders should communicate key messages
down to the individual level, but, in general, leaders and Soldiers should be able to tell
their stories unconstrained by overly prescriptive themes. When communicating, speed is
critical—minutes and hours matter—and we should remember to communicate to local
(Arabic/Iraqi) audiences first—U.S./global audience can follow.* Tell the truth, stay in
your lane, and get the message out fast. Be forthright and never allow enemy lies to stand
unchallenged. Demand accuracy, adequate context, and proper characterization from the
media.

Multinational Corps–Iraq
Counterinsurgency Guidance 2007

INFORMATION SUPERIORITY AND FULL SPECTRUM OPERATIONS

7-1.   Since Sun Tzu, successful rulers and military commanders have understood that knowing more than
their opponent and then acting on that knowledge effectively is critical. They also have understood the importance of persuasion in deciding which side ultimately prevails in any conflict. Conflicts in 21st century
conflicts occur in an operational environment of instant communications. Information systems are every-
where, exposure to news and opinion media is pervasive, the pace of change is increasing, and individual
actions can have immediate strategic implications. At every level—from the U.S. Government and its strategic communication, through joint capabilities used to exploit and degrade enemy command and control,
down to small-unit leaders meeting with village leaders—information shapes the operational environment.
It is a critical, and sometimes the decisive, factor in campaigns and major operations. Effectively employed, information multiplies the effects of friendly successes. Mishandled or ignored, it can lead to devastating reversals.

7-2.   *Information superiority* is the operational advantage derived from the ability to collect, process, and
disseminate an uninterrupted flow of information while exploiting or denying an adversary's ability to do
the same (JP 3-13). Innovative commanders drive their units to capitalize on the synergy between informational and other operational activities. They look to create and exploit opportunities to leverage the potential of both and to turn tactical events into opportunities for operational and strategic success. Information
is commanders' business. Commanders at every level require and use information to seize, retain, and exploit the initiative and achieve decisive results. Therefore, commanders must understand it, integrating it in
full spectrum operations as carefully as fires, maneuver, protection, and sustainment. Operations in Afghanistan illustrate this new reality.

7-3.   In January of 2007, a large Taliban force attempted to destroy a U.S. combat outpost near Margah in
the Afghan Province of Paktika. Seasoned by months of experience, the U.S. brigade combat team in that
area had organized their entire counterinsurgency operation around influencing specific audiences with
carefully combined information and action. The brigade identified, engaged, and destroyed the enemy
force as it moved into the area from Pakistan. In the ensuing week, with joint support, the brigade implemented a comprehensive information engagement plan to—

- Persuade the Afghan elders around Margah to deny support to the Taliban.
- Erode the cohesion, morale, and support base of the Taliban.

- Reassure the local population in Paktika Province.
- Persuade the Pakistani Army to take more active measures in Pakistan to disrupt the Taliban.

Additionally, the joint commander wanted to use this battle and other events to inform regional and global audiences about progress in this part of Afghanistan.

7-4. Soldiers gathered evidence and met with the local populace to ensure they understood the situation. The provincial reconstruction team helped the Afghan governor to organize a meeting with the Margah elders to pressure them into cutting ties with the Taliban. The attached psychological operations detachment developed and disseminated sophisticated products, targeting Taliban survivors of the battle. The public affairs officer then organized a press conference on-site in Margah to allow the Afghan governor to tell the story of the security success to local and regional audiences. The joint public affairs team organized a similar event for the international media. The joint commander met with senior commanders of the Pakistani and Afghan military.

7-5. The operation proved successful. The Pakistani Army improved security cooperation along the border. The Margah elders began to severe ties with the Taliban. Perhaps most importantly, the tribes in Pakistan began to resist Taliban recruiting efforts. Closely integrated information and action on the ground allowed joint and multinational forces to exploit tactical success.

7-6. Today, the struggle for information superiority takes place across many networks and in multiple domains. It impacts things as widely different as platforms in space, personal data assistants (known as PDAs), and the six o'clock news. It uses weapons that depend on advanced information technology for their devastating effectiveness, and it uses crude slogans and graffiti. U.S. forces have become the most sophisticated and powerful in the world by integrating information technology. Nonetheless, that very sophistication can make U.S. forces vulnerable to exploitation by an adversary. Exploitation ranges from sophisticated computer network attacks fully backed by a hostile power to an asymmetric blend of fanaticism, cell phones, garage door openers, messengers, and high-yield explosives. To counter these threats and focus on various audiences, commanders understand, visualize, describe, and direct efforts that contribute to information superiority. These contributors fall into four primary areas:

- Army information tasks—tasks used to shape the operational environment.
- Intelligence, surveillance, and reconnaissance—activities conducted to develop knowledge about the operational environment.
- Knowledge management—the art of using information to increase knowledge.
- Information management—the science of using information systems and methods.

ARMY INFORMATION TASKS

7-7. Full spectrum operations in today's operational environment require a comprehensive approach to information. In particular, Army operations emphasize the importance of peoples' perceptions, beliefs, and behavior to the success or failure of full spectrum operations and in the persistent conflicts the Nation continues to face. Enemies and adversaries will oppose friendly forces using every available information means, and they will exploit every advantage relentlessly. Army forces must protect friendly information and attack the opponents by using Army and joint capabilities. The Army conducts five information tasks to shape the operational environment. These are information engagement, command and control warfare, information protection, operations security, and military deception. (See table 7-1.)

7-8. Today's operational environment yields a high and often decisive impact to the side which best leverages information. As a result, commanders provide personal leadership, direction, and attention to it, fully integrating information into battle command. They integrate information tasks into all operations and include them in the operations process from inception. They incorporate cultural awareness, relevant social and political factors, and other informational aspects related to their mission in their understanding and visualization of the end state and operational design. In their guidance and intent, commanders clarify the effects they intend to achieve by synchronizing information and other operational activities, to include identifying relevant audiences and the desired perceptual or behavioral effects on each. Commanders match information tasks with actions on the ground in their concept of operations. They ensure their staffs incorporate implementing actions into tasks to subordinate units, coordinating instructions, and other parts of plans and orders. Commanders acquire the best situational understanding through visiting units and talking with Soldiers and others involved in operations. They capitalize on this knowledge to adjust actions and information tasks to gain the desired effects. Finally, commanders understand the advantages of building partner capacity in this critical mission area; they promote informational activity and capability by, with, and through host-nation forces.

Table 7-1. Army information tasks

Task	Information Engagement	Command and Control Warfare	Information Protection	Operations Security	Military Deception
Intended Effects	•Inform and educate internal and external publics •Influence the behavior of target audiences	•Degrade, disrupt, destroy, and exploit enemy command and control	•Protect friendly computer networks and communication means	•Deny vital intelligence on friendly forces to hostile collection	•Confuse enemy decision-makers
Capabilities	•Leader and Soldier engagement •Public affairs •Psychological operations •Combat camera •Strategic Communication and Defense Support to Public Diplomacy	•Physical attack •Electronic attack •Electronic warfare support •Computer network attack •Computer network exploitation	•Information assurance •Computer network defense •Electronic protection	•Operations security •Physical security •Counterintelligence	•Military deception

7-9.   The potential of information to generate powerful and perhaps unintended consequences can create a climate where risk aversion dominates decisionmaking related to information tasks. As a result, words and actions can fail to complement one another because there is no message, because the message is so neutral that it becomes irrelevant, or because the decision to employ a nonlethal capability is delayed until an opportunity is lost. Commanders overcome any such risk-averse tendencies by providing a clear, actionable, and achievable intent. They ensure that the timely and creative execution of information tasks is unhampered by overly cautious approval and control procedures.

INFORMATION ENGAGEMENT

7-10.   Land operations occur among populations. This requires Army forces to contend constantly with the attitudes and perceptions of populations within and beyond their area of operations. Commanders use information engagement in their areas of operation to communicate information, build trust and confidence, promote support for Army operations, and influence perceptions and behavior. ***Information engagement* is the integrated employment of public affairs to inform U.S. and friendly audiences; psychological operations, combat camera, U.S. Government strategic communication and defense support to public diplomacy, and other means necessary to influence foreign audiences; and, leader and Soldier engagements to support both efforts.** Commanders focus their information engagement activities on achieving desired effects locally. However, because land operations always take place in a broader global and regional context, commanders ensure their information engagement plans support and complement those of their higher headquarters, U.S. Government strategic communication guidance when available, and broader U.S. Government policy where applicable.

7-11. Soldiers' actions are the most powerful component of information engagement. Visible actions coordinated with carefully chosen, truthful words influence audiences more than either does alone. Local and regional audiences as well as adversaries compare the friendly force's message with its actions. People measure what they see and what they experience against the commander's messages. Consistency contributes to the success of friendly operations. Conversely, if actions and messages are inconsistent, friendly forces lose credibility. Loss of credibility makes land forces vulnerable to enemy and adversary actions and places Army forces at a disadvantage. Synchronizing information engagement with the overall operation ensures the messages are consistent with the force's actions and actions amplify the credibility of those messages.

Leader and Soldier Engagement

7-12. Face-to-face interaction by leaders and Soldiers strongly influences the perceptions of the local populace. Carried out with discipline and professionalism, day-to-day interaction of Soldiers with the local populace among whom they operate has positive effects. Such interaction amplifies positive actions, counters enemy propaganda, and increases goodwill and support for the friendly mission. Likewise, meetings conducted by leaders with key communicators, civilian leaders, or others whose perceptions, decisions, and actions will affect mission accomplishment can be critical to mission success. These meetings provide the most convincing venue for conveying positive information, assuaging fears, and refuting rumors, lies, and misinformation. Conducted with detailed preparation and planning, both activities often prove crucial in garnering local support for Army operations, providing an opportunity for persuasion, and reducing friction and mistrust.

Public Affairs

7-13. Public affairs is a commander's responsibility to execute public information, command information, and community engagement directed toward both the external and internal publics with interest in the Department of Defense.

7-14. Public affairs proactively informs and educates internal and external publics through public information, command information, and direct community engagement. Although all information engagement activities are completely truthful, public affairs is unique. It has a statutory responsibility to factually and accurately inform various publics without intent to propagandize or manipulate public opinion. Specifically, public affairs facilitates the commander's obligation to support informed U.S. citizenry, U.S. Government decisionmakers, and as operational requirements may dictate, non-U.S. audiences. Effective information engagement requires particular attention to clearly demarking this unique role of public affairs by protecting its credibility. This requires care and consideration when synchronizing public affairs with other information engagement activities. Public affairs and other information engagement tasks must be synchronized to ensure consistency, command credibility, and operations security.

7-15. The public affairs staff performs the following:—
- Advising and counseling the commander concerning public affairs.
- Public affairs planning.
- Media facilitation.
- Public affairs training.
- Community engagement.
- Communication strategies.

The public affairs staff requires augmentation to provide full support during protracted operations. (JP 3-61, AR 360-1, and FMs 46-1 and 3-61.1 govern public affairs.)

Psychological Operations

7-16. *Psychological operations* are planned operations to convey selected information and indicators to foreign audiences to influence their emotions, motives, objective reasoning, and ultimately the behavior of foreign governments, organizations, groups, and individuals. The purpose of psychological operations is to induce or reinforce foreign attitudes and behavior favorable to the originator's objectives (JP 1-02).

Commanders focus psychological operations efforts toward adversaries, their supporters, and their potential supporters. They may integrate these capabilities into the operations process through information engagement and the targeting process. Psychological operations units may also be task-organized with maneuver forces.

Combat Camera

7-17. *Combat camera* is the acquisition and utilization of still and motion imagery in support of combat, information, humanitarian, special force, intelligence, reconnaissance, engineering, legal, public affairs, and other operations involving the Military Services (JP 3-61). Combat camera generates still and video imagery in support of military operations. Combat camera units provide powerful documentary tools that support leader and Soldier engagement, psychological operations, and public affairs. For example, combat camera units can prepare products documenting Army tactical successes that counter enemy propaganda claiming the opposite.

Strategic Communication and Defense Support to Public Diplomacy

7-18. *Strategic communication* is focused United States Government efforts to understand and engage key audiences to create, strengthen, or preserve conditions favorable for the advancement of United States Government interests, policies, and objectives through the use of coordinated programs, plans, themes, messages, and products synchronized with the actions of all instruments of national power (JP 1-02). Strategic communication comprises an important part of the U.S. government's information arsenal. The government communicates themes and messages based on fundamental positions enumerated in the U.S. Constitution and further developed in U.S. policy. While U.S. leaders communicate some of this information directly through policy and directives, they also shape the environment by providing access and information to the media.

7-19. *Defense support to public diplomacy* is those activities and measures taken by the Department of Defense components to support and facilitate public diplomacy efforts of the United States Government (JP 3-13). Defense support to public diplomacy is a key military role in supporting the U.S. government's strategic communication program. It includes peacetime military engagement activities conducted as part of combatant commanders' theater security cooperation plans.

7-20. The Army implements strategic communication and defense support to public diplomacy while applying focused efforts to understand and engage key audiences. Such actions promote awareness, understanding, commitment, and action in support of the Army and its operations.

Responsibilities for Information Engagement

7-21. Commanders incorporate information engagement into full spectrum operations to impose their will on the operational environment. This requires commanders to be culturally astute; well-informed on the local political, social, and economic situations; and committed to leading the information engagement effort. Commanders direct multiple information engagement capabilities at those who affect or are affected by their operations. While doing so, they remain aware that, in an operational environment with pervasive connectivity and media presence, all messages ultimately reach all audiences. Spillover of a message intended for one audience to unintended audiences is inevitable. Dealing with the differing perspectives of diverse audiences requires thorough planning and continuously updated intelligence. Conducting full spectrum operations in the information age requires an accurate, complete, and clear understanding of each audience, including its interests, objectives, culture, and other nuances. Commanders reduce the natural ambiguity associated with this critical mission area by providing clear, actionable, and achievable intent and guidance.

7-22. Commanders integrate into information engagement into the operations process from inception, nesting information engagement activities with the intent of higher headquarters and with any applicable strategic communication guidance. They synchronize these activities with all other operational activity and integrate them into the operations process from inception. Finally, to prevent unintended consequences, commanders consider how actions proposed by the various staffs may affect the diverse audiences in the operational environment and their information engagement plan.

COMMAND AND CONTROL WARFARE

7-23. Information technology is becoming universally available. Most adversaries rely on communications and computer networks to make and implement decisions. Radios remain the backbone of tactical military command and control architectures. However, most communications relayed over radio networks are becoming digital as more computers link networks through transmitted frequencies. Additionally, adversaries are using civilian telecommunications, particularly cell phones and computer networks (including the Internet) to gather intelligence, disseminate information, shape perceptions and direct operations.

7-24. *Command and control warfare is* **the integrated use of physical attack, electronic warfare, and computer network operations, supported by intelligence, to degrade, destroy, and exploit the adversary's command and control system or to deny information to it.** It includes operations intended to degrade, destroy, and exploit an adversary's ability to use the electromagnetic spectrum and computer and telecommunications networks. These networks affect the adversary's command and control or ability to communicate with an external audience. Command and control warfare combines lethal and nonlethal actions. These actions degrade or destroy enemy information and the enemy's ability to collect and use that information. The fires cell synchronizes physical attack, electronic warfare, and computer network operations against enemy and adversary command and control.

7-25. Physical attack disrupts, damages, or destroys adversary targets through destructive combat power. In support of command and control warfare, it uses lethal action to destroy or degrade enemy command and control. The most common form of attack is through fires, although the targeting cell may develop priorities that require ground maneuver or aviation attack. Synchronizing physical attack, electronic attack, and computer network attack through the targeting process and integrating them into operations is fundamental to successful command and control warfare.

7-26. *Electronic warfare* is any military action involving the use of electromagnetic and directed energy to control the electromagnetic spectrum or to attack the enemy. Electronic warfare consists of three divisions: electronic attack, electronic protection, and electronic warfare support (JP 3-13.1). Of these, electronic attack and electronic exploitation directly support command and control warfare.

7-27. *Electronic attack* is that division of electronic warfare involving the use of electromagnetic energy, directed energy, or antiradiation weapons to attack personnel, facilities, or equipment with the intent of degrading, neutralizing, or destroying enemy combat capability and is considered a form of fires (JP 3-13.1).

7-28. *Electronic warfare support* is that division of electronic warfare involving actions tasked by, or under direct control of, an operational commander to search for, intercept, identify, and locate or localize sources of intentional and unintentional radiated electromagnetic energy for the purpose of immediate threat recognition, targeting, planning and conduct of future operations (JP 3-13.1).

7-29. *Computer network operations* are operations comprised of computer network attack, computer network defense, and related computer network exploitation enabling operations (JP 3-13). Of these, computer network attack and computer network exploitation directly support command and control warfare. *Computer network attack* consists of actions taken through the use of computer networks to disrupt, deny, degrade, or destroy information resident in computers and computer networks, or the computers and networks themselves (JP 3-13*). Computer network exploitation* are enabling operations and intelligence collection capabilities conducted through the use of computer networks to gather data from target or adversary automated information systems or networks (JP 3-13). Computer network defense is discussed under the information protection task. (See paragraph 7-33.)

7-30. Commanders use command and control warfare capabilities against an adversary's entire command and control system, not just the system's technical components. Although command and control warfare is primarily accomplished with physical and technical means, psychological operations and military deception activities can also provide important support, depending on the mission.

INFORMATION PROTECTION

7-31. *Information protection* **is active or passive measures that protect and defend friendly information and information systems to ensure timely, accurate, and relevant friendly information. It denies enemies, adversaries, and others the opportunity to exploit friendly information and information systems for their own purposes**. The secure and uninterrupted flow of data and information allows Army forces to multiply their combat power and sychronize landpower with other joint capabilities. Numerous threats to that capability exist in the operational environment. Information protection includes information assurance, computer network defense, and electronic protection. All three are interrelated.

7-32. *Information assurance* consists of measures that protect and defend information and information systems by ensuring their availability, integrity, authentication, confidentiality, and nonrepudiation. This includes providing for restoration of information systems by incorporating protection, detection, and reaction capabilities (JP 3-13).

7-33. *Computer network defense* consists of actions taken to protect, monitor, analyze, detect, and respond to unauthorized activity within the Department of Defense information systems and computer networks (JP 6-0). Effective network defense assures Army computer networks' functionality. It detects and defeats intruders attempting to exploit Army information and information systems. Commanders and staffs remain aware of and account for information on regulated (Department of Defense) and nonregulated (Internet) networks. They analyze how information from these mediums affect their operation; they take action to mitigate the associated risks.

7-34. *Electronic protection* is that division of electronic warfare involving actions taken to protect personnel, facilities, and equipment from any effects of friendly or enemy use of the electromagnetic spectrum that degrade, neutralize, or destroy friendly combat capability (JP 3-13.1).

OPERATIONS SECURITY

7-35. Operations security identifies essential elements of friendly information and evaluates the risk of compromise if an adversary or enemy obtains that information. This analysis compares the capabilities of hostile intelligence systems with the activities and communications of friendly forces and friendly information vulnerabilities. The analysis focuses on critical information that an adversary could interpret or piece together in time to be useful. Once identified, operations security experts prioritize friendly vulnerabilities and recommend countermeasures and other means of reducing the vulnerability. In some cases, the countermeasure cannot eliminate the risk, but it may reduce it to an acceptable level. Operations security includes physical security and counterintelligence. Physical security safeguards personnel, equipment, and information by preventing unauthorized access to equipment, installations, materiel, and documents while safeguarding them against espionage, sabotage, damage, and theft. Counterintelligence uses a wide range of information collection and activities to protect against espionage, combat other intelligence activities, protect against sabotage, and prevent assassinations. (JP 3-13.3 and FM 3-13 contain operations security doctrine.)

7-36. Operations security contributes to achieving surprise and completing the mission with little or no loss. Its absence contributes to excessive friendly casualties and possible mission failure. Information superiority hinges in no small part on effective operations security; therefore, measures to protect essential elements of friendly information cannot be an afterthought.

MILITARY DECEPTION

7-37. Military deception includes all actions conducted to mislead an enemy commander deliberately as to friendly military capabilities, intentions, and operations. At its most successful, military deception provokes an enemy commander to commit a serious mistake that friendly forces can exploit, there or elsewhere. However, effective military deception also introduces uncertainty into the enemy's estimate of the situation, and that doubt can lead to hesitation. Deception is a good means of dislocating an enemy force in time and space. Military deception can contribute significantly to information superiority; however, it requires integration into the overall operation beginning with receipt of mission. To achieve maximum effects, deceptions require good operations security, significant preparation, and resources for maximum effect. If added as an afterthought, deception often proves ineffective. Successful deception requires a reasonably accurate assessment of the enemy's expectations. (JP 3-13.4 and FM 3-13 contain military deception doctrine.)

INTELLIGENCE, SURVEILLANCE, AND RECONNAISSANCE

7-38. Knowledge of the operational environment is the precursor to all effective action, whether in the information or physical domain. Knowledge about the operational environment requires aggressive and continuous surveillance and reconnaissance to acquire information. Information collected from multiple sources and analyzed becomes intelligence that provides answers to commanders' information requirements concerning the enemy and other adversaries, climate, weather, terrain, and population. Developing this is the function of intelligence, reconnaissance and surveillance (ISR). *Intelligence, surveillance, and reconnaissance* **is an activity that synchronizes and integrates the planning and operation of sensors, assets, and processing, exploitation, and dissemination systems in direct support of current and future operations. This is an integrated intelligence and operations function. For Army forces, this activity is a combined arms operation that focuses on priority intelligence requirements while answering the commander's critical information requirements**. (JP 2-01 contains ISR doctrine.) Through ISR, commanders and staffs continuously plan, task, and employ collection assets and forces. These collect, process, and disseminate timely and accurate information, combat information, and intelligence to satisfy the commander's critical information requirements (CCIR) and other intelligence requirements. When necessary, ISR assets may focus on special requirements, such as information required for personnel recovery operations. It supports full spectrum operations through four tasks:

- ISR synchronization.
- ISR integration.
- Surveillance.
- Reconnaissance.

7-39. ISR synchronization considers all assets—both internal and external to the organization. It identifies information gaps and the most appropriate assets for collecting information to fill them. It also assigns the most efficient means to process the information into intelligence and disseminate it. ISR integration tasks assets to collect on requirements that intelligence reach or requests for information cannot answer or that commanders consider critical. Commanders integrate assets into a single ISR plan that capitalizes on each asset's capabilities. Commanders also synchronize and coordinate surveillance and reconnaissance missions and employ other units for ISR within the scheme of maneuver. Effectively synchronizing ISR with the overall plan positions ISR assets to continue to collect information, reconstitute for branches or sequels, or shift priorities throughout the operation.

INTELLIGENCE, SURVEILLANCE, AND RECONNAISSANCE SYNCHRONIZATION

7-40. *Intelligence, surveillance, and reconnaissance synchronization* **is the task that accomplishes the following: analyzes information requirements and intelligence gaps; evaluates available assets internal and external to the organization; determines gaps in the use of those assets; recommends intelligence, surveillance, and reconnaissance assets controlled by the organization to collect on the commander's critical information requirements; and submits requests for information for adjacent and higher collection support**. This task ensures that ISR, intelligence reach, and requests for information result in successful reporting, production, and dissemination of information, combat information, and intelligence to support decisionmaking.

7-41. The intelligence officer, with the operations officer and other staff elements, synchronizes the entire collection effort. This effort includes recommending tasking for assets the commander controls and submitting requests for information to adjacent and higher echelon units and organizations. When these sources do not answer the CCIR and other requirements, ISR synchronization uses intelligence reach to obtain the information.

7-42. ISR synchronization includes screening subordinate and adjacent unit requests for information concerning the enemy, terrain and weather, and civil considerations. When intelligence reach and requests for information do not satisfy a requirement, ISR synchronization develops specific information requirements to facilitate ISR integration. (FM 2-0 discusses intelligence reach.)

7-43. ISR synchronization is continuous. Commanders use it to assess ISR asset reporting. ISR synchronization includes continually identifying new and partially filled intelligence gaps. It also provides recommendations to the operations officer for tasking ISR assets.

INTELLIGENCE, SURVEILLANCE, AND RECONNAISSANCE INTEGRATION

7-44. *Intelligence, surveillance, and reconnaissance integration* **is the task of assigning and controlling a unit's intelligence, surveillance, and reconnaissance assets (in terms of space, time, and purpose) to collect and report information as a concerted and integrated portion of operation plans and orders**. This task ensures assignment of the best ISR assets through a deliberate and coordinated effort of the entire staff across all warfighting functions by integrating ISR into the operation.

7-45. The operations officer, with input from the intelligence officer, develops tasks based on specific information requirements (developed as part of ISR synchronization). Specific information requirements facilitate tasking by matching requirements to assets. The operations officer assigns tasks based on latest time that information is of value and the capabilities and limitations of available ISR assets. Intelligence requirements are identified, prioritized, and validated. An ISR plan is developed and synchronized with the overall operation. During ISR integration, the entire staff participates as responsibility for the ISR plan transitions from the intelligence officer to the operations officer. ISR integration is vital in controlling limited ISR assets. During ISR integration, the staff recommends redundancy and mix as appropriate. ISR synchronization and integration results in an effort focused on answering the commander's requirements through ISR tasks translated into orders.

SURVEILLANCE

7-46. *Surveillance* is the systematic observation of aerospace, surface, or subsurface areas, places, persons, or things, by visual, aural, electronic, photographic, or other means (JP 1-02). Surveillance involves observing an area to collect information.

7-47. Wide-area and focused surveillance missions provide valuable information. National and joint surveillance systems focus on information requirements for combatant commanders. They also provide information to all Services for operations across the area of responsibility. The systematic observation of geographic locations, persons, networks, or equipment is assigned to Army intelligence, reconnaissance, and maneuver assets. Changes or anomalies detected during surveillance missions can generate a reconnaissance mission to confirm or deny the change.

RECONNAISSANCE

7-48. *Reconnaissance* is a mission undertaken to obtain, by visual observation or other detection methods, information about the activities and resources of an enemy or adversary, or to secure data concerning the meteorological, hydrographic, or geographic characteristics of a particular area (JP 2-0).

7-49. Units performing reconnaissance collect information to confirm or deny current intelligence or predictions. This information may concern the terrain, weather, and population characteristics of a particular area as well the enemy. Reconnaissance normally precedes execution of the overall operation and extends throughout the area of operations. It begins as early as the situation, political direction, and rules of engagement permit. Reconnaissance can locate mobile enemy command and control assets—such as command posts, communications nodes, and satellite terminals—for neutralization, attack, or destruction. Reconnaissance can detect patterns of behavior exhibited by people in the objective area. Commanders at all echelons incorporate reconnaissance into their operations.

SOLDIER SURVEILLANCE AND RECONNAISSANCE

7-50. Surveillance is distinct from reconnaissance. Often surveillance is passive and may be continuous; reconnaissance missions are typically shorter and use active means (such as maneuver). Additionally, reconnaissance may involve fighting for information. Sometimes these operations are deliberate, as in a reconnaissance in force; however, the purpose of reconnaissance is to collect information, not initiate combat. Reconnaissance involves many tactics, techniques, and procedures throughout the course of a mission. An extended period of surveillance may be one of these. Commanders complement surveillance with frequent reconnaissance. Surveillance, in turn, increases the efficiency of reconnaissance by focusing those missions while reducing the risk to Soldiers.

7-51. The Soldier is an indispensable source for much of what the intelligence commanders need. Every Soldier is a sensor. Observations and experiences of Soldiers—who often work with the local populace—provide depth and context to information gathered through surveillance and reconnaissance. Commanders should train all Soldiers to report their observations, even when not assigned a surveillance or reconnaissance mission. Commanders and staffs emphasize integrating information gathered from Soldiers into intelligence production.

KNOWLEDGE AND INFORMATION MANAGEMENT

7-52. To respond to a rapidly changing operational environment and develop creativity, innovation, and adaptation, information must become knowledge. That knowledge must permeate throughout the Army. This requires both art and science. Knowledge management is the art of gaining and applying information throughout the Army and across the joint force. It generates knowledge products and services by and among commanders and staffs. It supports collaboration and the conduct of operations while improving organizational performance. Information management is the science of getting the right information to the right place in a timely manner and in such a way that it is immediately useable. It combines information systems and information processes to distribute, store, display, and protect knowledge products and services.

KNOWLEDGE MANAGEMENT

7-53. **Knowledge management is the art of creating, organizing, applying, and transferring knowledge to facilitate situational understanding and decisionmaking. Knowledge management supports improving organizational learning, innovation, and performance. Knowledge management processes ensure that knowledge products and services are relevant, accurate, timely, and useable to commanders and decisionmakers**. Knowledge management has three major components:

- **People**—those inside and outside the organization who create, organize, share, and use knowledge, and the leaders who foster an adaptive, learning environment.
- **Processes**—the methods to create, capture, organize, and apply knowledge.
- **Technology**—information systems that help collect, process, store, and display knowledge. Technology helps put knowledge products and services into organized frameworks.

7-54. Knowledge management exists to help commanders make informed, timely decisions despite the fog and friction of operations. It also enables effective collaboration by linking organizations and Soldiers requiring knowledge. Knowledge management enhances rapid adaptation in dynamic operations. It applies analysis and evaluation to information to create knowledge. Since a wide range of knowledge might affect operations, the commander's information requirements may extend beyond military matters. Defining these requirements is an important aspect of knowledge management. Establishing their CCIRs is one way commanders define their information requirements. The CCIRs focus development of knowledge products.

7-55. All leaders need to understand the processes and procedures associated with the systems available to share information and acquire knowledge. Commanders and staffs assess knowledge management effectiveness by considering whether it lessens the fog of war. Knowledge management narrows the gap between relevant information commanders require and that which they have. Developing a knowledge management plan is necessary to accomplish the following:

- Address knowledge and information flow.
- Develop criteria for displaying the common operational picture.
- Access and filter information from sources normally found outside the military or the organization.
- Support developing situational awareness and situational understanding.
- Enable rapid, accurate retrieval of previously developed knowledge to satisfy new requirements.
- Route products to the appropriate individuals in a readily understood format.
- Keep commanders and staffs from being overwhelmed by information.

7-56. Staff responsibility for knowledge management begins with the chief of staff. Depending on the complexity of the situation, it may require dedicated personnel. Effective knowledge management requires effective information management.

Situational Awareness

7-57. Commanders begin their visualization with situational awareness. **Situational awareness is immediate knowledge of the conditions of the operation, constrained geographically and in time.** More simply, it is Soldiers knowing what is currently happening around them. Situational awareness occurs in Soldiers' minds. It is not a display or the common operational picture; it is the interpretation of displays or the actual observation of a situation. On receipt of mission, commanders develop their situational awareness. They base it on information and knowledge products, such as the common operational picture and running estimates.

Situational Understanding

7-58. During mission analysis, commanders apply judgment to their situational awareness to arrive at situational understanding. **Situational understanding is the product of applying analysis and judgment to relevant information to determine the relationships among the mission variables to facilitate decisionmaking.** It enables commanders to determine the implications of what is happening and forecast what may happen. Situational understanding enhances decisionmaking by identifying opportunities, threats to the force or mission accomplishment, and information gaps. It helps commanders identify enemy options and likely future actions, the probable consequences of proposed friendly actions, and the effect of the operational environment on both. Situational understanding based on a continuously updated common operational picture fosters individual initiative by reducing, although not eliminating, uncertainty.

INFORMATION MANAGEMENT

7-59. **Information management is the science of using procedures and information systems to collect, process, store, display, disseminate, and protect knowledge products, data, and information.** Information management disseminates timely and protected relevant information to commanders and staffs. Information management helps commanders develop situational understanding. It also helps them make and disseminate effective decisions faster than the enemy can. Among other aspects, information management includes lower level mechanical methods, such as organizing, collating, plotting, and arranging. However, information management is more than technical control of data flowing across networks. It employs both staff management and automatic processes to focus a vast array of information and make relevant information available to the right person at the right time. Information management centers on commanders and the information they need to exercise command and control. It has two components: information systems and relevant information.

Information Systems

7-60. An **information system is equipment and facilities that collect, process, store, display, and disseminate information. This includes computers—hardware and software—and communications, as well as policies and procedures for their use.** Information systems are the physical dimension of information management. They employ automated processes that sort, filter, store, and disseminate information according to the commander's priorities. These capabilities relieve the staff of handling routine data. Effective information systems automatically process, disseminate, and display information according to user requirements. Information systems, especially when merged into a single, integrated network, enable extensive information sharing. Commanders make the best use of information systems when they determine their information requirements and focus their staffs and organizations on meeting them.

7-61. LandWarNet is the Army's portion of the Global Information Grid. LandWarNet encompasses all Army information management systems and information systems that collect, process, store, display, disseminate, and protect information worldwide. It enables execution of Army command and control processes and supports operations through wide dissemination of relevant information. LandWarNet facilitates rapidly converting relevant information into decisions and actions. It allows commanders to exercise command and control from anywhere in their area of operations. (JP 6-0 describes the Global Information Grid.)

Relevant Information

7-62. *Relevant information* **is all information of importance to commanders and staffs in the exercise of command and control.** To be relevant, information must be accurate, timely, usable, complete, precise, reliable, and secure. Relevant information provides the answers commanders and staffs need to conduct operations successfully. The mission variables are the categories of relevant information. (FM 6-0 contains doctrine on relevant information and the cognitive hierarchy. The cognitive hierarchy describes how data becomes information, knowledge, and understanding.)

7-63. Effective information management identifies relevant information and processes data into information for development into and use as knowledge. Information management then quickly routes information and knowledge products to those who need them. All information given to commanders should be relevant information. That is, commanders should only receive information or knowledge products that they need for exercising command and control. The information commanders receive drives how they visualize the operation. How relevant information fits into the commander's visualization determines its value. In turn, their visualization guides what information commanders seek. Commanders emphasize the most important relevant information they need by establishing CCIRs. Providing the information commanders need to make decisions and maintain an accurate situational understanding requires staffs to understand the commander's intent and know the CCIRs.

Information Categories

7-64. Information management places information into one of four categories: specified requirements, implied requirements, gaps, and distractions. Specified requirements are requirements commanders specifically identify. CCIRs, priority intelligence requirements, and friendly force information requirements are categories of specified requirements. Implied requirements are important pieces of information that commanders need but have not requested. Effective staffs develop implied requirements and recommend them for specified requirements. These often become priority intelligence requirements or friendly force information requirements. Gaps are elements of information commanders need to achieve situational understanding but do not have. Ideally, analysis identifies gaps and translates them into specified requirements. ISR focuses on collecting information to fill gaps. Until a gap is filled, commanders and staffs make assumptions, clearly identifying them as such. This practice is most common during planning. Staffs continually review assumptions and replace them with facts as information becomes available. Distractions include information commanders do not need to know but continue to receive. Distractions contribute to information overload.

7-65. Effective information management keeps commanders and staffs aware of the quality of their information as they use it to build situational understanding. Soldiers processing information use these criteria to evaluate the quality of an element of information:

- **Relevance**—applies to the mission, situation, or task at hand.
- **Accuracy**—conveys the true situation.
- **Timeliness**—is available in time to make decisions.

- **Usability**—is portrayed in common, easily understood formats and displays.
- **Completeness**—provides all necessary data.
- **Precision**—has the required level of detail.
- **Security**—affords required protection.

SUMMARY

7-66. Mission success in the information age demands land forces possess unparalleled ability to use information to seize, retain, and exploit the initiative and to achieve decisive results. It demands that commanders prevail in a continuous struggle for information superiority, a struggle that begins before deploying forces and continues long after concluding traditional military activities. To that end, commanders use ISR, coupled with knowledge management and information management, to make better decisions more rapidly than their enemies and adversaries. Simultaneously, they direct information tasks to hamper their opponents' decisionmaking ability, protect their own, gain the trust and confidence of the people, and win the support of the diverse audiences throughout their operational environment. Finally, they ensure efforts to gain information superiority are synchronized into the operations process from inception, integrating information, a critical element of combat power, into their exercise of battle command.

This page intentionally left blank.

Chapter 8

Strategic and Operational Reach

Army forces require strategic and operational reach to deploy and immediately conduct operations anywhere with little or no advanced notice. Contemporary operations require Army forces that can deploy rapidly and conduct extended campaigns. These operations require Soldiers and units with campaign and expeditionary capabilities. Commanders and organizations require proficiency at force projection, protection, and sustainment. Soldiers require an expeditionary mindset to prepare them for short-notice deployments into uncertain, often austere environments. This chapter discusses how strategic and operational reach affects deploying and employing Army forces. It also addresses principles to maximize the effects of both factors. Strategic and operational reach depend on basing in and near the joint operations area.

STRATEGIC REACH

8-1.   Strategic reach provides the capability to operate against complex, adaptive threats operating anywhere. The distance across which the Nation can project decisive military power is its strategic reach. This multifaceted reach combines joint military capabilities—air, land, maritime, space, special operations, and cyber—with those of the other instruments of national power. Land force capabilities complement those of other Services. Army forces increase the joint force's strategic reach by securing and operating bases far from the United States. However, Army forces depend on joint-enabled force projection capabilities to deploy and sustain them across intercontinental distances. In some cases, land forces use strategic lift to deploy directly to an operational area. In many instances, land operations combine direct deployment with movements from intermediate staging bases located outside the operational area. Access to bases and support depends upon the Nation's diplomatic and economic power as well as its military capabilities.

OPERATIONAL MANEUVER FROM STRATEGIC DISTANCE

8-2.   Operational maneuver from strategic distance combines global force projection with maneuver against an operationally significant objective. (See figure 8-1, page 8-2.) It requires strategic reach that deploys maneuverable landpower to an operational area in a position of advantage. It requires enough operational reach to execute operations decisively without an operational pause. It aims to avoid operational pauses associated with various requirements. Then it can secure and defend a lodgment; develop a base; and receive, stage, and build up forces. Success demands full integration of all available joint means. Thus, it combines force projection with land maneuver to operational depth in an integrated, continuous operation.

8-3.   The most difficult form of operational maneuver from strategic distance projects forces directly from the United States into an operational area. Examples of this operational maneuver from strategic distance include the 1942 invasion of North Africa and the 1992 intervention in Somalia. These operations involved forces projected from the United States with near simultaneous employment. In many cases, operational maneuver from strategic distance requires intermediate staging bases. From these bases, operational maneuver develops using intratheater lift and Army maneuver capabilities. The availability of bases in a region extends the strategic reach of Army forces; bases near the operational area increase opportunities for successful operational maneuver from strategic distance.

8-4.   Today, joint forces combine strategic and operational reach in forcible and unopposed entry operations. These operations originate from outside the operational area, often using intermediate staging bases.

Entry operations conducted across intercontinental distances capitalize on the U.S. dominance of the air and sea. Exploiting these capabilities creates a dilemma for opponents.

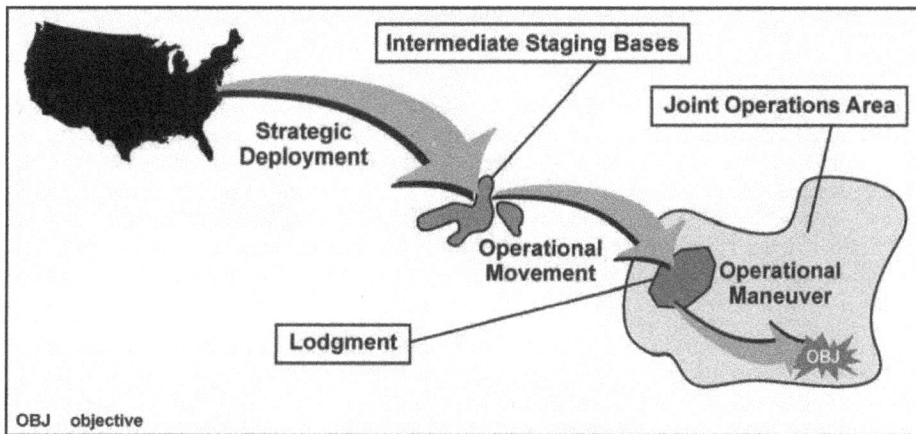

Figure 8-1. Operational maneuver from strategic distance

EXPEDITIONARY CAMPAIGNS

8-5. Expeditionary campaigns are inherently joint operations founded in strategic reach. During crisis response, joint force commanders rely on contingency expeditionary forces to respond promptly. The Army provides ready forces able to operate in any environment—from urban areas to remote, rural regions. The early introduction of credible, capable forces into the operational area is an important strategic factor. This action may quickly convince a potential enemy that further aggression would be too costly. Initial-entry forces need to be interoperable and flexible enough to contend with unforeseen circumstances. Immediately upon arrival, initial-entry forces require enough combat power to establish and protect lodgments and begin simultaneous shaping operations. The ability to fight at the outset is crucial to successfully executing the campaign plan. A tailored force able to dominate situations early enables the joint force commander to seize, retain, and exploit the initiative.

8-6. Expeditionary forces are configured for immediate employment in austere conditions. They do not depend on existing infrastructure; they adapt as the situation evolves. Commanders prepare to transition to sustained operations, assimilate new capabilities into force packages for follow-on operations, or disperse until otherwise required. As a result, the force composition varies throughout the campaign based on the dominant element of full spectrum operations.

8-7. If campaign objectives are not achieved swiftly, the Army provides forces tailored with the combat power and endurance needed for sustained, joint-enabled land operations. Endurance gives the force the ability to employ combat power anywhere for protracted periods. It allows Army forces to preserve the gains of initial operations and complements and reinforces efforts of the other instruments of national power to achieve strategic objectives. It has strategic, operational, and tactical implications. Strategically, endurance requires the Army to sustain the generation and rotation of forces sufficient to meet combatant commanders' requirements. At the operational level, endurance requires Army forces with enough operational reach to complement other joint capabilities throughout a campaign. Tactically, endurance is a function of sustainment and protection. Endurance gives Army forces their campaign capability.

EXPEDITIONARY ARMY FORCES

8-8. Expeditionary forces impose a distinct set of dynamics on Army forces. The Army depends on strategic deployment assets provided by U.S. Transportation Command. Combatant commanders establish priorities to move forces into the operational area. That decision drives allocation of strategic lift and ultimately determines how rapidly Army forces deploy. Although U.S. strategic lift assets exceed those of any other nation, the available lift is rarely sufficient to deploy a large force at one time. Consequently, commanders carefully select elements of the force and the sequence in which they deploy to match operational area conditions. This is force tailoring.

8-9.   The range of possible employment options complicates training. Army forces cannot train for every possible mission; they train for full spectrum operations with emphasis on the most likely mix of tasks. When designated as an expeditionary force for specific missions, Army forces modify training and prepare for that contingency. The volatile nature of crises requires Army forces to simultaneously train, deploy, and execute. Commanders conduct operations with initial-entry forces while assembling and preparing follow-on forces. Commanders carefully consider and accept risk to create opportunity. Such actions help them to seize the initiative during deployment and in the early phases of an operation, even when the situation is not fully developed. Balancing these dynamics is an art mastered through knowledge, experience, and judgment.

8-10. The payoff for mastering expeditionary operations is mission accomplishment. Fast deploying and expansible Army forces give joint force commanders the means to introduce operationally significant land forces into a crisis on short notice. These expeditionary capabilities arm joint force commanders with a preemptive ability to deter, shape, fight, and win. Rapidly deployed expeditionary force packages provide immediate options for seizing or retaining the operational initiative. They complement and reinforce the other Services with modular capability packages that can be swiftly tailored, deployed, and employed.

8-11. Expeditionary capability is more than the ability to deploy quickly. It requires deploying the right mix Army forces to the right place at the right time. It provides joint force commanders with the ability to deter an adversary or take decisive action if deterrence fails. Forward deployed units, forward positioned capabilities, peacetime military engagement, and force projection—from anywhere in the world—all contribute to expeditionary capabilities.

8-12. When deployed, every unit regardless of type generates combat power and contributes to the success of the mission. From the operational and tactical perspectives, commanders ensure deployed Army forces possess sufficient combat power to overwhelm any potential enemy. The art of expeditionary warfare balances the ability to mass the effects of lethal combat systems against the requirements to deploy, support, and sustain units that use those systems. Commanders assemble force packages that maximize the lethality of initial-entry forces. These packages remain consistent with both the mission and the requirement to project, employ, and sustain the force. Commanders tailor follow-on forces to increase both the lethality and operational reach of the entire force.

8-13. Deploying commanders integrate protection capabilities to ensure mission accomplishment and increase the survivability of deployed Army forces. As with the other attributes, lift constraints and time available complicate the situation. Survivability requires astutely assessing operational risk. In many operations, rapid offensive action to seize the initiative may better protect forces than massive defenses around lodgments.

8-14. Generating and sustaining combat power is fundamental to expeditionary warfare. Commanders reconcile competing requirements. Army forces must accomplish joint force commander-assigned missions, yet they require adequate sustainment for operations extended in time and space. Commanders tailor force packages to provide sufficient sustainment capability without amassing a significant sustainment footprint. Wherever possible, commanders augment sustainment capacity with host-nation and contracted support.

FORCE PROJECTION

8-15. Force projection is the military component of power projection. It is a central element of the National Military Strategy. Speed is paramount; force projection is a race between friendly forces and the enemy or situation. The side that achieves an operational capability first can seize the initiative. Thus, it is not the velocity of individual stages or transportation means that is decisive; it is a combat-ready force deployed to an operational area before the enemy is ready or the situation deteriorates further.

8-16. Commanders visualize force projection as one seamless operation. Deployment speed sets the initial tempo of military activity in the operational area. Commanders understand how speed, sequence, and mix of deploying forces affect their employment options. They see how their employment concept establishes deployment requirements. Commanders prioritize the mix of forces on the time-phased force and deployment list to project forces into the operational area where and when required. Singular focus on the land component, to the exclusion of complementary joint capabilities, may result in incorrect force sequencing. Commanders exercise active and continuous command and control during force projection. They couple it with detailed reverse-sequence planning. Thus, the right forces and the right support are available and ready to conduct decisive operations when needed.

8-17. Force projection encompasses five processes: mobilization, deployment, employment, sustainment, and redeployment. These processes occur in a continuous, overlapping, and repeating sequence throughout an operation. Force projection operations are inherently joint. They require detailed planning and synchronization. Sound, informed decisions made early on force projection may determine a campaign's success.

8-18. Each process has its own criteria. Mobilization is the process of bringing the armed forces to a state of readiness in response to a contingency. Deployment is the relocation of forces and materiel to a desired operational area in response to a contingency. It has four supporting components: predeployment activities; fort to port; port to port; and reception, staging, onward movement, and integration. Employment is the conduct of operations to support a joint force commander. Sustainment involves providing and maintaining personnel and materiel required to support a joint force commander. Redeployment is the return of forces and materiel to the home or mobilization station. (JP 4-05 discusses mobilization. JP 3-0 discusses employment. JP 4-0 and FM 4-0 discuss sustainment. FMI 3-35 discusses deployment and redeployment of brigade-sized and larger Army forces. FM 4-01.011 discusses deployment and redeployment of battalion-sized and smaller Army forces.)

8-19. Reception, staging, onward movement, and integration focuses on reassembling deploying units and quickly integrating them into the force. It is the critical link between deploying and employing forces. Effective reception, staging, onward movement, and integration establishes a smooth flow of personnel, equipment, and materiel from ports of debarkation through employment as reassembled, mission-capable forces. A deploying unit is most vulnerable between its arrival and operational employment, so protection is vital.

8-20. Closing the force more rapidly can be achieved by increasing the allocation of strategic lift or by using pre-positioned equipment sets. (FMs 100-17-1 and 100-17-2 discuss pre-positioned operations.) Combatant commanders and Army Service component commands can also facilitate force projection through anticipatory actions. Actions may include positioning equipment or troops in anticipated crisis areas, securing access to ports and airfields, enhancing capabilities of regional forces, and protecting areas critical to force projection.

ENTRY OPERATIONS

8-21. Whenever possible, Army forces seek an unopposed entry, either unassisted or assisted by the host nation. An assisted entry requires host-nation cooperation. In an assisted entry, initial-entry forces are tailored to deploy efficiently and transition quickly to follow-on operations. Reception, staging, onward movement, and integration focuses on cooperating with the host nation to expedite moving units from ports of debarkation to tactical assembly areas. In an unassisted entry, no secure facilities for deploying forces exist. The joint force commander deploys balanced force packages with enough combat power to secure an adequate lodgment and perform reception, staging, onward movement, and integration. Force sequencing for an unassisted entry is similar to that of a forcible entry.

8-22. A *forcible entry* is the seizing and holding of a military lodgment in the face of armed opposition (JP 3-18). Once the assault force seizes the lodgment, it normally defends to retain it while the joint force commander rapidly deploys additional combat power by air and sea. When conditions are favorable, joint force commanders may combine a forcible entry with other offensive operations in a coup de main. This action can achieve the strategic objectives in a simultaneous major operation. The 1989 invasion of

Panama is an example of operational maneuver from strategic distance in a coup de main. (JP 3-18 contains joint doctrine for forcible entry operations.)

8-23. A forcible entry operation can be by parachute, air, or amphibious assault. The Army's parachute assault and air assault forces provide a formidable forcible entry capability. Marine forces specialize in amphibious assault; they also conduct air assaults as part of amphibious operations. Special operations forces play an important role in forcible entry; they conduct shaping operations in support of conventional forces while executing their own missions. These capabilities permit joint force commanders to overwhelm enemy anti-access measures and quickly insert combat power. The entry force either resolves the situation or secures a lodgment for delivery of larger forces by aircraft or ships. The three forms of forcible entry produce complementary and reinforcing effects that help joint force commanders to seize the initiative early in a campaign.

8-24. Forcible entry operations are inherently complex and always joint. Often only hours separate the alert from deployment. The demands of simultaneous deployment and employment create a distinct set of dynamics. Operations are carefully planned and rehearsed in training areas and marshalling areas. Personnel and equipment are configured for employment upon arrival without reception, staging, onward movement, and integration.

OPERATIONAL REACH

8-25. The challenge of conducting and sustaining operations over long distances has been studied and theorized upon since the time of Sun Tzu. History is replete with examples of campaigns plagued with inadequate operational reach. To achieve the desired end state, forces must possess the necessary operational reach to establish and maintain conditions that define success. (See paragraphs 6-74 through 6-79.)

8-26. Extending operational reach is a paramount concern for commanders. Commanders and staffs increase operational reach through deliberate, focused operational design. Operational design balances the natural tension between tempo, endurance, and risk to increase operational reach. A well-designed operation, executed skillfully, extends operational reach several ways, to include—

- Setting the tempo of the operation for greater endurance.
- Phasing the operation to assure its continuation.
- Employing the support of other Service components to relieve land forces of tasks that detract from the decisive operation.

BASING

8-27. A *base* is a locality from which operations are projected or supported (JP 1-02). The base includes installations and facilities that provide sustainment. Bases may be joint or single Service areas. Commanders often designate a specific area as a base and assign responsibility for protection and terrain management within the base to a single commander. Units located within the base are under the tactical control of the base commander. Within large bases, controlling commanders may designate base clusters for mutual protection and command and control.

8-28. Strategic and operational reach initially depend upon basing in the area of responsibility and overflight rights. Both affect how much combat power can be generated in the operational area in a prescribed period. The arrangement and location of forward bases (often in austere, rapidly emplaced configurations) complement the ability of Army forces to conduct sustained, continuous combat operations to operational depth. Though typically determined by diplomatic and political considerations, basing and overflight rights are essential to the commander's ability to maintain or extend operational reach.

8-29. Army forces typically rely on a mix of intermediate staging bases, lodgments (subsequently developed into bases), and forward operating bases to deploy and employ landpower simultaneously to operational depth. These bases establish and maintain strategic reach for deploying forces and ensure sufficient operational reach to extend operations in time and space.

INTERMEDIATE STAGING BASES

8-30. An *intermediate staging base* is a tailorable, temporary location used for staging forces, sustainment and/or extraction into and out of an operational area (JP 3-35). At the intermediate staging base, units are unloaded from intertheater lift, reassembled and integrated with their equipment, and then moved by intratheater lift into the operational area. The theater army commander provides extensive support to Army forces transiting the base. The combatant commander may designate the theater army commander to command the base or provide a headquarters suitable for the task. Intermediate staging bases are established near, but normally not in, the operational area. They often are located in the supported combatant commander's area of responsibility. For land forces, intermediate staging bases may be located in the operational area. However, they are always established outside the range of enemy fires and beyond the enemy's political sphere of influence.

8-31. Ideally, secure bases will be available in the operational area, but conditions that compel deployment may also negate the availability of a secure lodgment in the operational area. Under these circumstances, an intermediate staging base may serve as the principal staging base for entry operations. In cases where the force needs to secure a lodgment, an intermediate staging base may be critical to success.

8-32. Normally, intermediate staging bases exploit advantages of existing, developed capabilities, serving as efficient transfer points from high-volume commercial carriers to various tactical, intratheater transport means. However, these bases are transshipment points. Using them can increase handling requirements and deployment times and may require infrastructure development to support further deployment. When deciding whether to operate through an intermediate staging base, commanders carefully weigh advantages gained by deploying through the base against operational risks, such as time, lift, and distance, associated with its utilization.

LODGMENTS

8-33. A *lodgment* is a designated area in a hostile or potentially hostile territory that, when seized and held, makes the continuous landing of troops and materiel possible and provides maneuver space for subsequent operations (JP 3-18). Identifying and preparing the initial lodgment significantly influences the conduct of an operation. Lodgments should expand to allow easy access to strategic sealift and airlift, offer adequate space for storage, facilitate transshipment of supplies and equipment, and be accessible to multiple lines of communications. Typically, deploying forces establish lodgments near key points of entry in the operational area that offer central access to air, land, and sea transportation hubs.

8-34. A lodgment rarely possesses the ideal characteristics desired to support ongoing operations. Improving the base capabilities may require early deployment of maintenance, engineering, or terminal operations forces. Contracting, medical, legal, and financial management personnel who arrange access to host-nation capabilities should be among the first to deploy. The requirement for adequate sustainment capability is especially critical in the operation's early stages when building combat power is critical and forces are most vulnerable. Identifying infrastructure requirements during mission analysis is essential to establishing the lodgment and enhancing the responsiveness and sustainability of the force.

8-35. The time required to establish a lodgment depends on the extent and condition of the civil and military infrastructure present in the operational area. In areas where extensive industrial facilities and distribution capabilities exist and are available, commanders can initiate operations without a significant pause. In the absence of these capabilities, the force cannot begin operations until a sufficient base is established and operational. In more austere environments, where initial entry and operations may be severely restricted, acquisition, construction, and sustainment capabilities should arrive early in the deployment flow. This arrival improves the lodgment, generates and moves forces and materiel into forward operating bases, and establishes operational-level sustainment capability to support the deployed force.

FORWARD OPERATING BASES

8-36. A *forward operating base* is an area used to support tactical operations without establishing full support facilities. Such bases may be used for an extended time. During protracted operations, they may be expanded and improved to establish a more permanent presence. The scale and complexity of the base, however, directly relates to the size of the force required to maintain it. A large base with extensive facilities requires a much larger security force than a smaller, austere base. Commanders weigh whether to expand and improve a forward operating base against the type and number of forces available to secure it, the expected length of the forward deployment, and the force's sustainment requirements.

8-37. Forward operating bases extend and maintain the operational reach of Army forces, providing secure locations from which to conduct and sustain operations. They not only enable extending operations in time and space; they also contribute to the overall endurance of the force, an essential element of the Army's campaign capability. Forward operating bases allow forward-deployed forces to reduce operational risk, maintain momentum, and avoid culmination.

8-38. Typically, forward operating bases are established adjacent to a regional distribution hub, such as a large airfield (civilian or military), rail terminal, or major highway junction. This facilitates movement into and out of the operational area while providing a secure location through which to distribute personnel, equipment, and supplies. However, forward operating bases may be located in austere locations with limited access to transportation infrastructure. In such cases, maintaining the base for extended periods is unlikely.

SUMMARY

8-39. The Nation requires joint forces with strategic and operational reach. Given the enormous distances that separate the United States from regions in conflict, this imposes serious challenges for the Army. Even within the United States, the distances between Army installations and major cities may be significant. Above all, the Army must be versatile, adapting not only to the particular requirements of different areas of responsibility but also to limitations in strategic and intratheater lift. Available lift will never equal an ideal land force's requirements. Joint force commanders need some landpower deployed very rapidly and capable of seizing a lodgment. They also need follow-on land forces able to persevere for months and years as the campaign progresses. Once deployed, Army commanders develop and protect bases and lines of communications in austere areas. These house not only Soldiers but also joint and multinational forces. With each base, Army forces extend their operational reach throughout the operational area, using landpower to multiply the effectiveness of American military power.

This page intentionally left blank.

Appendix A

Principles of War and Operations

The nine principles of war represent the most important nonphysical factors that affect the conduct of operations at the strategic, operational, and tactical levels. The Army published its original principles of war after World War I. In the following years, the Army adjusted the original principles modestly as they stood the tests of analysis, experimentation, and practice. The principles of war are not a checklist. While they are considered in all operations, they do not apply in the same way to every situation. Rather, they summarize characteristics of successful operations. Their greatest value lies in the education of the military professional. Applied to the study of past campaigns, major operations, battles, and engagements, the principles of war are powerful analysis tools. Joint doctrine adds three principles of operations to the traditional nine principles of war.

OBJECTIVE

Direct every military operation toward a clearly defined, decisive, and attainable objective.

A-1. The principle of objective drives all military activity. At the operational and tactical levels, objective ensures all actions contribute to the higher commander's end state. When undertaking any mission, commanders should clearly understand the expected outcome and its impact. Combat power is limited; commanders never have enough to address every aspect of the situation. Objectives allow commanders to focus combat power on the most important tasks. Clearly stated objectives also promote individual initiative. These objectives clarify what subordinates need to accomplish by emphasizing the outcome rather than the method. Commanders should avoid actions that do not contribute directly to achieving the objectives.

A-2. The purpose of military operations is to accomplish the military objectives that support achieving the conflict's overall political goals. In offensive and defensive operations, this involves destroying the enemy and his will to fight. The objective of stability or civil support operations may be more difficult to define; nonetheless, it too must be clear from the beginning. Objectives must contribute to the operation's purpose directly, quickly, and economically. Each tactical operation must contribute to achieving operational and strategic objectives.

A-3. Military leaders cannot dissociate objective from the related joint principles of restraint and legitimacy, particularly in stability operations. The amount of force used to obtain the objective must be prudent and appropriate to strategic aims. Means used to accomplish the military objective must not undermine the local population's willing acceptance of a lawfully constituted government. Without restraint or legitimacy, support for military action deteriorates, and the objective becomes unobtainable.

OFFENSIVE

Seize, retain, and exploit the initiative.

A-4. As a principle of war, offensive is synonymous with initiative. The surest way to achieve decisive results is to seize, retain, and exploit the initiative. Seizing the initiative dictates the nature, scope, and tempo of an operation. Seizing the initiative compels an enemy to react. Commanders use initiative to impose their will on an enemy or adversary or to control a situation. Seizing, retaining, and exploiting the initiative are all essential to maintain the freedom of action necessary to achieve success and exploit vulnerabilities. It helps commanders respond effectively to rapidly changing situations and unexpected developments.

A-5. In combat operations, offensive action is the most effective and decisive way to achieve a clearly defined objective. Offensive operations are the means by which a military force seizes and holds the initiative while maintaining freedom of action and achieving decisive results. The importance of offensive action is fundamentally true across all levels of war. Defensive operations shape for offensive operations by economizing forces and creating conditions suitable for counterattacks.

MASS

Concentrate the effects of combat power at the decisive place and time.

A-6. Commanders mass the effects of combat power in time and space to achieve both destructive and constructive results. Massing in time applies the elements of combat power against multiple decisive points simultaneously. Massing in space concentrates the effects of combat power against a single decisive point. Both can overwhelm opponents or dominate a situation. Commanders select the method that best fits the circumstances. Massed effects overwhelm the entire enemy or adversary force before it can react effectively.

A-7. Army forces can mass lethal and nonlethal effects quickly and across large distances. This does not imply that they accomplish their missions with massed fires alone. Swift and fluid maneuver based on situational understanding complements fires. Often, this combination in a single operation accomplishes what formerly took an entire campaign.

A-8. In combat, commanders mass the effects of combat power against a combination of elements critical to the enemy force to shatter its coherence. Some effects may be concentrated and vulnerable to operations that mass in both time and space. Other effects may be spread throughout depth of the operational area, vulnerable only to massing effects in time.

A-9. Mass applies equally in operations characterized by civil support or stability. Massing in a stability or civil support operation includes providing the proper forces at the right time and place to alleviate suffering and provide security. Commanders determine priorities among the elements of full spectrum operations and allocate the majority of their available forces to the most important tasks. They focus combat power to produce significant results quickly in specific areas, sequentially if necessary, rather than dispersing capabilities across wide areas and accomplishing less.

ECONOMY OF FORCE

Allocate minimum essential combat power to secondary efforts.

A-10. Economy of force is the reciprocal of mass. Commanders allocate only the minimum combat power necessary to shaping and sustaining operations so they can mass combat power for the decisive operation. This requires accepting prudent risk. Taking calculated risks is inherent in conflict. Commanders never leave any unit without a purpose. When the time comes to execute, all units should have tasks to perform.

MANEUVER

Place the enemy in a disadvantageous position through the flexible application of combat power.

A-11. Maneuver concentrates and disperses combat power to keep the enemy at a disadvantage. It achieves results that would otherwise be more costly. Effective maneuver keeps enemy forces off balance by making them confront new problems and new dangers faster than they can counter them. Army forces gain and preserve freedom of action, reduce vulnerability, and exploit success through maneuver. Maneuver is more than just fire and movement. It includes the dynamic, flexible application of all the elements of combat power. It requires flexibility in thought, plans, and operations. In operations dominated by stability or civil support, commanders use maneuver to interpose Army forces between the population and threats to security and to concentrate capabilities through movement.

UNITY OF COMMAND

For every objective, ensure unity of effort under one responsible commander.

A-12. Applying a force's full combat power requires unity of command. Unity of command means that a single commander directs and coordinates the actions of all forces toward a common objective. Cooperation may produce coordination, but giving a single commander the required authority is the most effective way to achieve unity of effort.

A-13. The joint, interagency, intergovernmental, and multinational nature of unified action creates situations where the commander does not directly control all organizations in the operational area. In the absence of command authority, commanders cooperate, negotiate, and build consensus to achieve unity of effort.

SECURITY

Never permit the enemy to acquire an unexpected advantage.

A-14. Security protects and preserves combat power. Security results from measures a command takes to protect itself from surprise, interference, sabotage, annoyance, and threat surveillance and reconnaissance. Military deception greatly enhances security.

SURPRISE

Strike the enemy at a time or place or in a manner for which he is unprepared.

A-15. Surprise is the reciprocal of security. It is a major contributor to achieving shock. It results from taking actions for which the enemy is unprepared. Surprise is a powerful but temporary combat multiplier. It is not essential to take enemy forces completely unaware; it is only necessary that they become aware too late to react effectively. Factors contributing to surprise include speed, operations security, and asymmetric capabilities.

SIMPLICITY

Prepare clear, uncomplicated plans and clear, concise orders to ensure thorough understanding.

A-16. Plans and orders should be simple and direct. Simple plans and clear, concise orders reduce misunderstanding and confusion. The situation determines the degree of simplicity required. Simple plans executed on time are better than detailed plans executed late. Commanders at all levels weigh potential benefits of a complex concept of operations against the risk that subordinates will fail to understand or follow it. Orders use clearly defined terms and graphics. Doing this conveys specific instructions to subordinates with reduced chances for misinterpretation and confusion.

A-17. Multinational operations put a premium on simplicity. Differences in language, doctrine, and culture complicate them. Simple plans and orders minimize the confusion inherent in this complex environment. The same applies to operations involving interagency and nongovernmental organizations.

ADDITIONAL PRINCIPLES OF JOINT OPERATIONS

A-18. In addition to these nine principles, JP 3-0 adds three principles of operations—perseverance, legitimacy, and restraint. Together with the principles of war, these twelve make up the principles of joint operations.

PERSEVERANCE

Ensure the commitment necessary to attain the national strategic end state.

A-19. Commanders prepare for measured, protracted military operations in pursuit of the desired national strategic end state. Some joint operations may require years to reach the desired end state. Resolving the underlying causes of the crisis may be elusive, making it difficult to achieve conditions supporting the end state. The patient, resolute, and persistent pursuit of national goals and objectives often is a requirement for success. This will frequently involve diplomatic, informational, and economic measures to supplement military efforts. In the end, the will of the American public, as expressed through their elected officials and advised by expert military judgment, determines the duration and size of any military commitment.

A-20. Army forces' endurance and commanders' perseverance are necessary to accomplish long-term missions. A decisive offensive operation may swiftly create conditions for short-term success. However, protracted stability operations, executed simultaneously with defensive and offensive tasks, may be needed to achieve the strategic end state. Commanders balance their desire to enter the operational area, accomplish the mission quickly, and depart against broader requirements. These include the long-term commitment needed to achieve national goals and objectives.

LEGITIMACY

Develop and maintain the will necessary to attain the national strategic end state.

A-21. For Army forces, legitimacy comes from three important factors. First, the operation or campaign must be conducted under U.S. law. Second, the operation must be conducted according to international laws and treaties recognized by the United States, particularly the law of war. Third, the campaign or operation should develop or reinforce the authority and acceptance for the host-nation government by both the governed and the international community. This last factor is frequently the decisive element.

A-22. Legitimacy is also based on the will of the American people to support the mission. The American people's perception of legitimacy is strengthened if obvious national or humanitarian interests are at stake. Their perception also depends on their assurance that American lives are not being placed at risk needlessly or carelessly.

A-23. Other interested audiences may include foreign nations, civil populations in and near the operational area, and participating multinational forces. Committed forces must sustain the legitimacy of the operation and of the host-nation government, where applicable. Security actions must balance with the need to maintain legitimacy. Commanders must consider all actions potentially competing for strategic and tactical requirements. All actions must exhibit fairness in dealing with competing factions where appropriate. Legitimacy depends on the level of consent to the force and to the host-nation government, the people's expectations, and the force's credibility.

RESTRAINT

Limit collateral damage and prevent the unnecessary use of force.

A-24. Restraint requires careful and disciplined balancing of security, the conduct of military operations, and the desired strategic end state. Excessive force antagonizes those friendly and neutral parties involved. Hence, it damages the legitimacy of the organization that uses it while potentially enhancing the legitimacy of any opposing party. The rules of engagement must be carefully matched to the strategic end state and the situation. Commanders at all levels ensure their personnel are properly trained in rules of engagement and quickly informed of any changes. Rules of engagement may vary according to national policy concerns but should always be consistent with the inherent right of self-defense.

A-25. Restraint is best achieved when rules of engagement issued at the beginning of an operation address a range of plausible situations. Commanders should consistently review and revise rules of engagement as necessary. Additionally, commanders should carefully examine them to ensure that the lives and health of Soldiers are not needlessly endangered. National concerns may lead to different rules of engagement for multinational participants; commanders must be aware of national restrictions imposed on force participants.

Appendix B

Command and Support Relationships

Command and support relationships provide the basis for unity of command and unity of effort in operations. Command relationships affect Army force generation, force tailoring, and task organization. Commanders use Army support relationships when task-organizing Army forces. All command and support relationships fall within the framework of joint doctrine. JP 1 discusses joint command relationships and authorities.

CHAIN OF COMMAND

B-1. The President and Secretary of Defense exercise authority and control of the armed forces through two distinct branches of the chain of command as described in JP 1. (See figure B-1 [taken from JP 1], page B-2.) One branch runs from the President, through the Secretary of Defense, to the combatant commanders for missions and forces assigned to combatant commands. The other branch runs from the President through the Secretary of Defense to the secretaries of the military departments. This branch is used for purposes other than operational direction of forces assigned to the combatant commands. Each military department operates under the authority, direction, and control of the secretary of that military department. These secretaries exercise authority through their respective Service chiefs over Service forces not assigned to combatant commanders. The Service chiefs, except as otherwise prescribed by law, perform their duties under the authority, direction, and control of the secretaries to whom they are directly responsible.

B-2. The typical operational chain of command extends from the combatant commander to a joint task force commander, then to a functional component commander or a Service component commander. Joint task forces and functional component commands, such as a land component, comprise forces that are normally subordinate to a Service component command but have been placed under the operational control (OPCON) of the joint task force, and subsequently to a functional component commander. Conversely, the combatant commander may designate one of the Service component commanders as the joint task force commander or as a functional component commander. In some cases, the combatant commander may not establish a joint task force, retaining operational control over subordinate functional commands and Service components directly.

B-3. Under joint doctrine, each joint force includes a Service component command that provides administrative and logistic support to Service forces under OPCON of that joint force. However, Army doctrine distinguishes between the Army component of a combatant command and Army components of subordinate joint forces. Under Army doctrine, Army Service component command (ASCC) refers to the Army component assigned to a combatant command. There is only one ASCC within a combatant command's area of responsibility. The Army components of all other joint forces are called ARFORs. **An *ARFOR* is the Army Service component headquarters for a joint task force or a joint and multinational force**. It consists of the senior Army headquarters and its commander (when not designated as the joint force commander) and all Army forces that the combatant commander subordinates to the joint task force or places under the control of a multinational force commander. The ARFOR becomes the conduit for most Service-related issues and administrative support. The Army Service component command may function as an ARFOR headquarters when the combatant commander does not exercise command and control through subordinate joint force commanders.

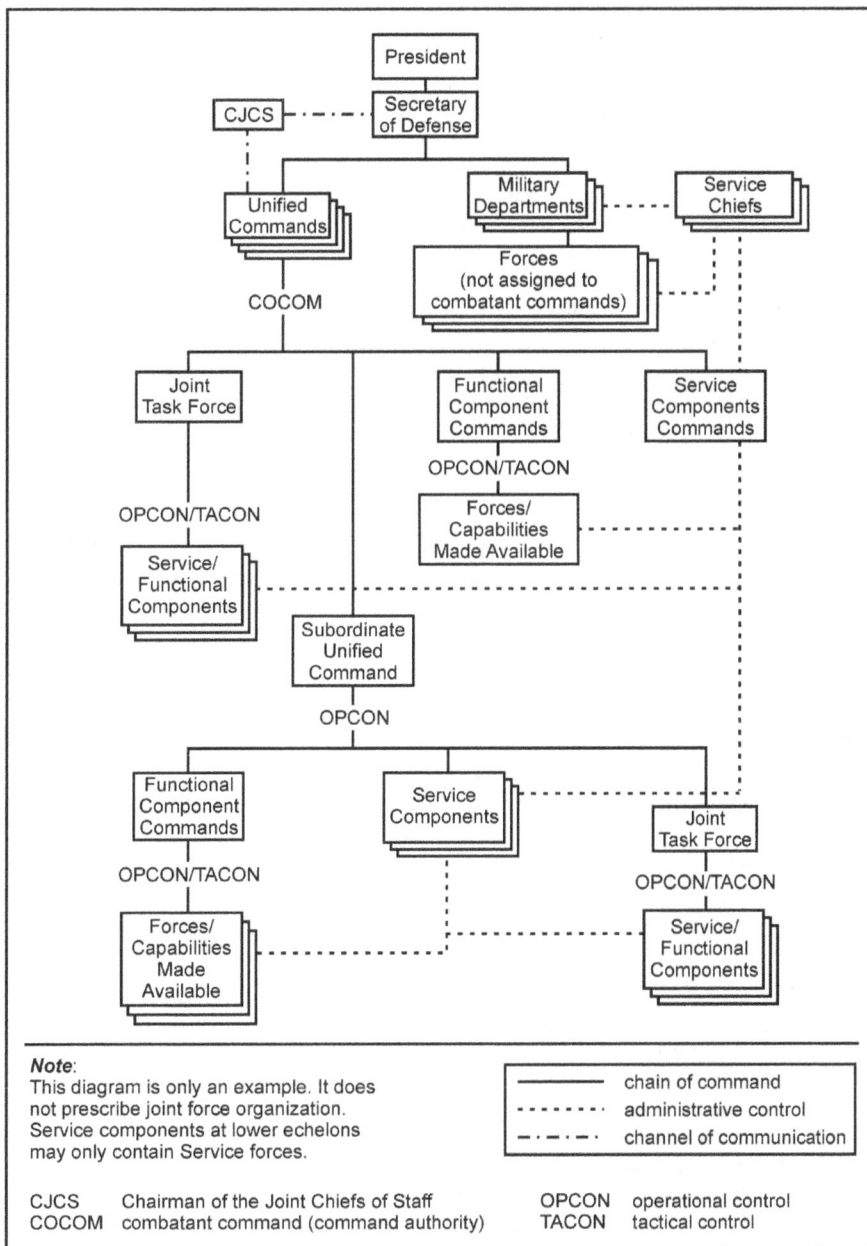

Figure B-1. Chain of command branches

B-4. The Secretary of the Army directs the flow of administrative control (ADCON). (See paragraphs B-25 through B-27.) Administrative control for Army units within a combatant command normally extends from the Secretary of the Army through the ASCC, through an ARFOR, and then to Army units assigned or attached to an Army headquarters within that joint command. However, administrative control is not tied to the operational chain of command. The Secretary of the Army may redirect some or all Service responsibilities outside the normal ASCC channels. In similar fashion, the ASCC may distribute some administrative responsibilities outside the ARFOR. Their primary considerations are the effectiveness of Army forces and the care of Soldiers.

COMBATANT COMMANDS

B-5. The Unified Command Plan establishes combatant commanders' missions and geographic responsibilities. Combatant commanders directly link operational military forces to the Secretary of Defense and the President. The Secretary of Defense deploys troops and exercises military power through the combatant commands. Six combatant commands have areas of responsibility. They are the geographic combatant commands. Each geographic combatant command has (or will have) an assigned ASCC. For doctrinal purposes, these commands become "theater armies" to distinguish them from the similar organizations assigned to functional component commands. The geographic combatant commands and their theater armies are—

- U.S. Northern Command (U.S. Army, North—USARNORTH).
- U.S. Southern Command (U.S. Army, South—USARSO).
- U.S. Central Command (U.S. Army, Central—USARCENT).
- U.S. European Command (U.S. Army, Europe—USAREUR).
- U.S. Pacific Command (U.S. Army, Pacific—USARPAC).
- U.S. Africa Command (U.S. Army, Africa—USARAF) implementation date pending.

In addition to these geographic combatant commands, U.S. Forces Korea is a subordinate unified command of U.S. Pacific Command. It has a theater army also (Eighth Army—EUSA).

B-6. There are four functional combatant commands. Each has global responsibilities. Like the geographic combatant commands, each has an ASCC assigned. These organizations are not theater armies; they are functional Service component commands. The functional combatant commands and their associated ASCCs are—

- U.S. Joint Forces Command (U.S. Army Forces Command—FORSCOM).
- U.S. Strategic Command (U.S. Army Space and Missile Defense Command/Army Strategic Command—SMDC/ARSTRAT).
- U.S. Special Operations Command (U.S. Army Special Operations Command—USASOC).
- U.S. Transportation Command (Military Surface Deployment and Distribution Command—SDDC).

JOINT TASK FORCES AND SERVICE COMPONENTS

B-7. Joint task forces are the organizations most often used by a combatant commander for contingencies. Combatant commanders establish joint task forces and designate the joint force commanders for these commands. Those commanders exercise OPCON of all U.S. forces through functional component commands, Service components, subordinate joint task forces, or a combination of these. (See figure B-2 [taken from JP 1], page B-4.) The senior Army officer assigned to a joint task force, other than the joint force commander and members of the joint task force staff, becomes the ARFOR commander. The ARFOR commander answers to the Secretary of the Army through the ASCC for most ADCON responsibilities.

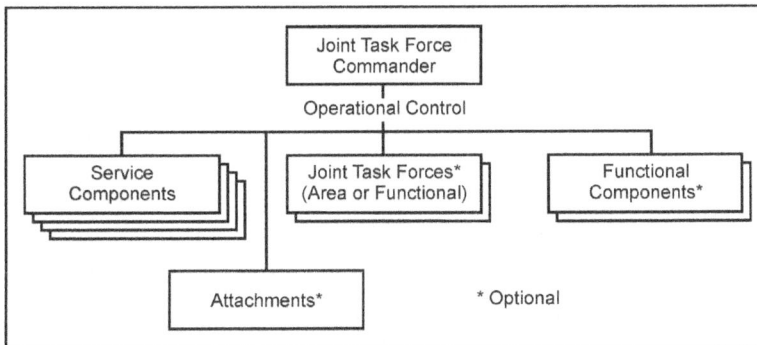

Figure B-2. Joint task force organization options

B-8. Depending on the joint task force organization, the ARFOR commander may exercise OPCON of some or all Army forces assigned to the task force. For example, an Army corps headquarters may become a joint force land component within a large joint task force. (See figure B-3, which shows an example of a joint task force organized into functional components.) The corps commander exercises OPCON of Army divisions and tactical control (TACON) of Marine Corps forces within the land component. As the senior Army headquarters, the corps becomes the ARFOR for not only the Army divisions but also all other Army units within the joint task force, including those not under OPCON of the corps. This ensures that Service responsibilities are fulfilled while giving the joint force commander maximum flexibility for employing the joint force. Unless modified by the Secretary of the Army or the ASCC, Service responsibilities continue through the ARFOR to the respective Army commanders. Army forces in figure B-3 are shaded to show this relationship. The corps has OPCON of the Army divisions and TACON of the Marine division. The corps does not have OPCON over the other Army units but does, as the ARFOR, exercise ADCON over them. The corps also assists the ASCC in controlling Army support to other Services and to any multinational forces as directed.

B-9. When an Army headquarters becomes the joint force land component as part of a joint task force, Army units subordinated to it are normally under OPCON. Marine Corps forces made available to a joint force land component command built around an Army headquarters are normally under TACON. The land component commander makes recommendations to the joint force commander on properly using attached, OPCON, or TACON assets; planning and coordinating land operations; and accomplishing such operational missions as assigned.

B-10. Navy and Coast Guard forces often operate under a joint force maritime component commander. This commander makes recommendations to the joint force commander on proper use of assets attached or under OPCON or TACON. Maritime component commanders also make recommendations concerning planning and coordinating maritime operations and accomplishing such missions.

B-11. A joint force air component commander is normally designated the supported commander for the air interdiction and counterair missions. The air component command is typically the headquarters with the majority of air assets. Like the other functional component commanders, the air component commander makes recommendations to the joint force commander on properly using assets attached or under OPCON or TACON. The air component commander makes recommendations for planning and coordinating air operations and accomplishing such missions. Additionally, the air component commander often has responsibility for airspace control authority and area air defense.

B-12. The joint force commander may organize special operations forces and conventional forces together as a joint special operations task force or a subordinate joint task force. Other functional components and subordinate task forces—such as a joint logistic task force or joint psychological operations task force— are established as required.

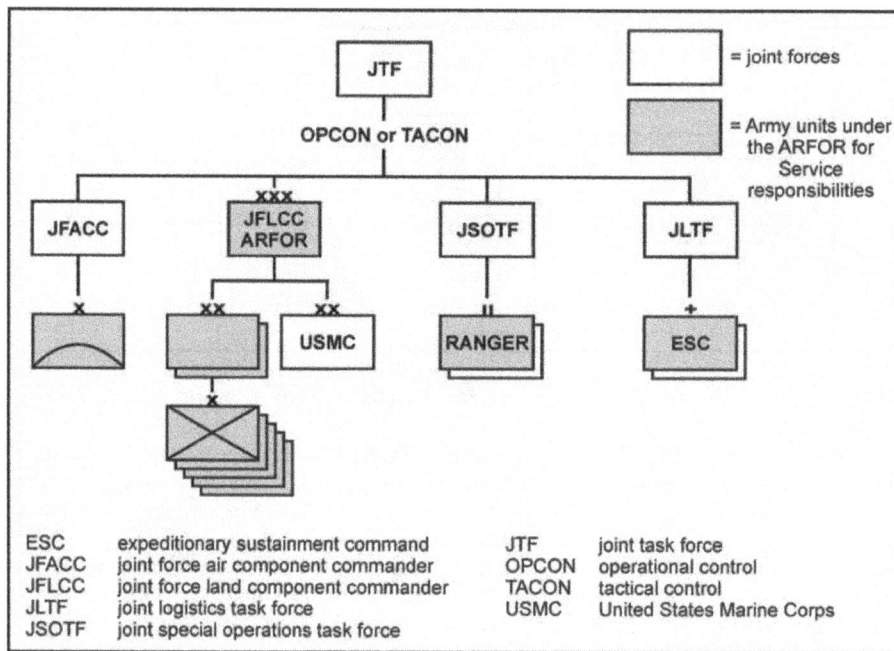

Figure B-3. Example of a joint task force showing an Army corps as joint force land component commander with ARFOR responsibilities

JOINT COMMAND RELATIONSHIPS

B-13. JP 1 specifies and details four types of joint command relationships:

- Combatant command (command authority) (COCOM).
- Operational control.
- Tactical control.
- Support.

The following paragraphs summarize important provisions of these relationships. The glossary contains complete definitions.

COMBATANT COMMAND (COMMAND AUTHORITY)

B-14. COCOM is the command authority over assigned forces vested only in commanders of combatant commands or as directed by the President or the Secretary of Defense in the Unified Command Plan and cannot be delegated or transferred. Title 10, U.S. Code, section 164 specifies it in law. Normally, the combatant commander exercises this authority through subordinate joint force commanders, Service component, and functional component commanders. COCOM includes the directive authority for logistic matters (or the authority to delegate it to a subordinate joint force commander for common support capabilities required to accomplish the subordinate's mission).

OPERATIONAL CONTROL

B-15. OPCON is the authority to perform those functions of command over subordinate forces involving—

- Organizing and employing commands and forces.
- Assigning tasks.
- Designating objectives.
- Giving authoritative direction necessary to accomplish missions.

B-16. OPCON normally includes authority over all aspects of operations and joint training necessary to accomplish missions. It does not include directive authority for logistics or matters of administration, discipline, internal organization, or unit training. The combatant commander must specifically delegate these elements of COCOM. OPCON does include the authority to delineate functional responsibilities and operational areas of subordinate joint force commanders. In two instances, the Secretary of Defense may specify adjustments to accommodate authorities beyond OPCON in an establishing directive: when transferring forces between combatant commanders or when transferring members and/or organizations from the military departments to a combatant command. Adjustments will be coordinated with the participating combatant commanders. (JP 1 discusses operational control in detail.)

TACTICAL CONTROL

B-17. TACON is inherent in OPCON. It may be delegated to and exercised by commanders at any echelon at or below the level of combatant command. TACON provides sufficient authority for controlling and directing the application of force or tactical use of combat support assets within the assigned mission or task. TACON does not provide organizational authority or authoritative direction for administrative and logistic support; the commander of the parent unit continues to exercise these authorities unless otherwise specified in the establishing directive. (JP 1 discusses tactical control in detail.)

SUPPORT

B-18. Support is a command authority in joint doctrine. A supported and supporting relationship is established by a superior commander between subordinate commanders when one organization should aid, protect, complement, or sustain another force. Designating supporting relationships is important. It conveys priorities to commanders and staffs planning or executing joint operations. Designating a support relationship does not provide authority to organize and employ commands and forces, nor does it include authoritative direction for administrative and logistic support. Joint doctrine divides support into the categories listed in table B-1.

Table B-1. Joint support categories

Category	Definition
General support	That support which is given to the supported force as a whole and not to any particular subdivision thereof (JP 1-02).
Mutual support	That support which units render each other against an enemy, because of their assigned tasks, their position relative to each other and to the enemy, and their inherent capabilities (JP 1-02).
Direct support	A mission requiring a force to support another specific force and authorizing it to answer directly to the supported force's request for assistance (JP 3-09.1).
Close support	That action of the supporting force against targets or objectives which are sufficiently near the supported force as to require detailed integration or coordination of the supporting action with the fire, movement, or other actions of the supported force (JP 1-02).

B-19. Support is, by design, somewhat vague but very flexible. Establishing authorities ensure both supported and supporting commanders understand the authority of supported commanders. Joint force commanders often establish supported and supporting relationships among components. For example, the maritime component commander is normally the supported commander for sea control operations; the air component commander is normally the supported commander for counterair operations. An Army head-quarters designated as the land component may be the supporting force during some campaign phases and the supported force in other phases.

B-20. The joint force commander may establish a support relationship between functional and Service component commanders. Conducting operations across a large operational area often involves both the land and air component commanders. The joint task force commander places the land component in general support of the air component until the latter achieves air superiority. Conversely, within the land area of operations, the land component commander becomes the supported commander and the air component commander provides close support. A joint support relationship is not used when an Army commander task-organizes Army forces in a supporting role. When task-organized to support another Army force, Army forces use one of four Army support relationships. (See paragraphs B-35 through B-36.)

JOINT ASSIGNMENT AND ATTACHMENT

B-21. All forces under the jurisdiction of the secretaries of the military departments (with exception) are assigned to combatant commands or the commander, U.S. Element North America Aerospace Defense Command (known as USELEMNORAD). The exception exempts those forces necessary to carry out the functions of the military departments as noted in Title 10, U.S. Code, section 162. The assignment of forces to the combatant commands comes from the Secretary of Defense in the "Forces for Unified Commands" memorandum. According to this memorandum and the Unified Command Plan, unless otherwise directed by the President or the Secretary of Defense, all forces operating in the geographic area assigned to a specific combatant commander are assigned or attached to that combatant commander. A force assigned or attached to a combatant command may be transferred from that command to another combatant commander only when directed by the Secretary of Defense and approved by the President. The Secretary of Defense specifies the command relationship the gaining commander will exercise (and the losing commander will relinquish). Establishing authorities for subordinate unified commands and joint task forces may direct the assignment or attachment of their forces to those subordinate commands and delegate the command relationship as appropriate. (See JP 1.)

B-22. When the Secretary of Defense assigns Army forces to a combatant command, the transfer is either permanent or the duration is unknown but very lengthy. The combatant commander exercises COCOM over assigned forces. When the Secretary of Defense attaches Army units, this indicates that the transfer of units is relatively temporary. Attached forces normally return to their parent combatant command at the end of the deployment. The combatant commander exercises OPCON of the attached force. In either case, the combatant commander normally exercises OPCON over Army forces through the ASCC until the combatant commander establishes a joint task force or functional component. At that time, the combatant commander delegates OPCON to the joint task force commander. When the joint force commander establishes any command relationship, the ASCC clearly specifies ADCON responsibilities for all affected Army commanders.

COORDINATING AUTHORITY

B-23. Coordinating authority is the authority delegated to a commander or individual for coordinating specific functions or activities involving forces of two or more military departments, two or more joint force components, or two or more forces of the same Service. The commander or individual granted coordinating authority can require consultation between the agencies involved but does not have the authority to compel agreement. In the event that essential agreement cannot be obtained, the matter shall be referred to the appointing authority. Coordinating authority is a consultation relationship, not an authority through which command may be exercised. Coordinating authority is more applicable to planning and similar activities than to operations. (See JP 1.) For example, a joint security commander exercises coordinating authority over area security operations within the joint security area. Commanders or leaders at any echelon at or below combatant command may be delegated coordinating authority. These individuals may be assigned responsibilities established through a memorandum of agreement between military and nonmilitary organizations.

DIRECT LIAISON AUTHORIZED

B-24. *Direct liaison authorized* is that authority granted by a commander (any level) to a subordinate to directly consult or coordinate an action with a command or agency within or outside of the granting command. Direct liaison authorized is more applicable to planning than operations and always carries with it the requirement of keeping the commander granting direct liaison authorized informed. Direct liaison authorized is a coordination relationship, not an authority through which command may be exercised (JP 1).

ADMINISTRATIVE CONTROL

B-25. *Administrative control* is direction or exercise of authority over subordinate or other organizations in respect to administration and support, including organization of Service forces, control of resources and equipment, personnel management, unit logistics, individual and unit training, readiness, mobilization, demobilization, discipline, and other matters not included in the operational missions of the subordinate or other organizations (JP 1). It is a Service authority, not a joint authority. It is exercised under the authority of and is delegated by the Secretary of the Army. ADCON is synonymous with the Army's Title 10 authorities and responsibilities.

B-26. ADCON of Army forces involves the entire Army. Figure B-4 identifies major responsibilities of the Department of the Army and illustrates their normal distribution between the Army generating force and operating forces. The generating force consists of those Army organizations whose primary mission is to generate and sustain the operational Army's capabilities for employment by joint force commanders. *Operating forces* consist of those forces whose primary missions are to participate in combat and the integral supporting elements thereof (JP 1-02). Often, commanders in the operating force and commanders in the generating force subdivide specific responsibilities. Army generating force capabilities and organizations are linked to operating forces through co-location and reachback.

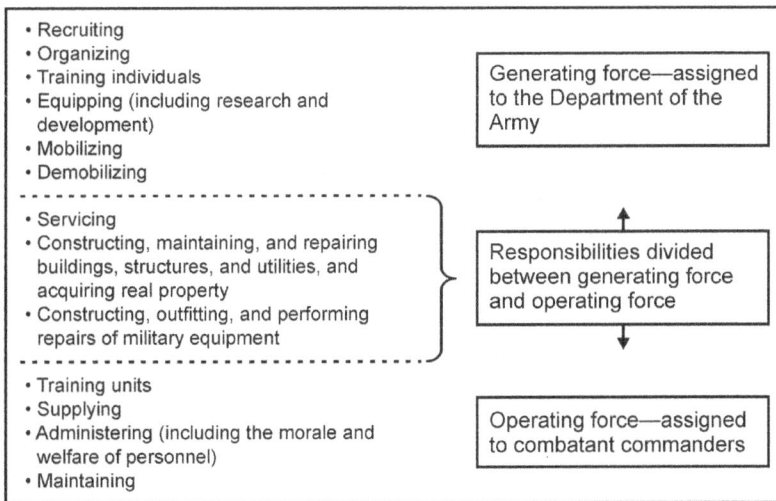

• Recruiting • Organizing • Training individuals • Equipping (including research and development) • Mobilizing • Demobilizing	Generating force—assigned to the Department of the Army

• Servicing
• Constructing, maintaining, and repairing
  buildings, structures, and utilities, and
  acquiring real property
• Constructing, outfitting, and performing
  repairs of military equipment
	Responsibilities divided
	between generating force
	and operating force

• Training units • Supplying • Administering (including the morale and welfare of personnel) • Maintaining	Operating force—assigned to combatant commanders

Figure B-4. Normal distribution of Army administrative control responsibilities

B-27. The ASCC is always the senior Army headquarters assigned to a combatant command. Its commander exercises command authorities as assigned by the combatant commander and ADCON as delegated by the Secretary of the Army. ADCON is the Army's authority to administer and support Army forces even while in a combatant command area of responsibility. COCOM is the basic authority for command and control of the same Army forces. The Army is obligated to meet the combatant commander's requirements for the operational forces. Essentially, ADCON directs the Army's support of operational force requirements. Unless modified by the Secretary of the Army, administrative responsibilities normally flow from Department of the Army through the ASCC to those Army forces assigned or attached to that combatant command. ASCCs usually "share" ADCON for at least some administrative or support functions. "Shared ADCON" refers to the internal allocation of Title 10, U.S. Code, section 3013(b) responsibilities and functions. This is especially true for Reserve Component forces. Certain administrative functions, such as pay, stay with the Reserve Component headquarters, even after unit mobilization. Shared ADCON also applies to direct reporting units of the Army that typically perform single or unique functions. The direct reporting unit, rather than the ASCC, typically manages individual and unit training for these units. The Secretary of the Army directs shared ADCON.

ARMY COMMAND AND SUPPORT RELATIONSHIPS

B-28. Army command relationships are similar but not identical to joint command authorities and relationships. Differences stem from the way Army forces task-organize internally and the need for a system of support relationships between Army forces. Another important difference is the requirement for Army commanders to handle the administrative support requirements that meet the needs of Soldiers. These differences allow for flexible allocation of Army capabilities within various Army echelons. Army command and support relationships are the basis for building Army task organizations. A *task organization* is a **temporary grouping of forces designed to accomplish a particular mission**. Certain responsibilities are inherent in the Army's command and support relationships.

ARMY COMMAND RELATIONSHIPS

B-29. Table B-2, page B-10, lists the Army command relationships. Command relationships define superior and subordinate relationships between unit commanders. By specifying a chain of command, command relationships unify effort and give commanders the ability to employ subordinate forces with maximum flexibility. Army command relationships identify the degree of control of the gaining Army commander. The type of command relationship often relates to the expected longevity of the relationship between the headquarters involved and quickly identifies the degree of support that the gaining and losing Army commanders provide.

B-30. *Organic* forces are those assigned to and forming an essential part of a military organization. Organic parts of a unit are those listed in its table of organization for the Army, Air Force, and Marine Corps, and are assigned to the administrative organizations of the operating forces for the Navy (JP 1-02). Joint command relationships do not include organic because a joint force commander is not responsible for the organizational structure of units. That is a Service responsibility.

B-31. The Army establishes organic command relationships through organizational documents such as tables of organization and equipment and tables of distribution and allowances. If temporarily task-organized with another headquarters, organic units return to the control of their organic headquarters after completing the mission. To illustrate, within a brigade combat team (BCT), the entire brigade is organic. In contrast, within most modular support brigades, there is a "base" of organic battalions and companies and a variable mix of assigned and attached battalions and companies. (See appendix C.)

B-32. Army assigned units remain subordinate to the higher headquarters for extended periods, typically years. Assignment is based on the needs of the Army and is formalized by orders rather than organizational documents. Although force tailoring or task-organizing may temporarily detach units, they eventually return to their either their headquarters of assignment or their organic headquarters. Attached units are temporarily subordinated to the gaining headquarters, and the period may be lengthy, often months or longer. They
return to their parent headquarters (assigned or organic) when the reason for the attachment ends. The Army headquarters that receives another Army unit through assignment or attachment assumes responsibility for the ADCON requirements, and particularly sustainment, that normally extend down to that echelon, unless modified by directives or orders. For example, when an Army division commander attaches an engineer battalion to a brigade combat team, the brigade commander assumes responsibility for the unit training, maintenance, resupply, and unit-level reporting for that battalion.

Table B-2. Command relationships

If relation-ship is:	Then inherent responsibilities:							
	Have command relation-ship with:	May be task-organized by:[1]	Unless modified, ADCON responsi-bility goes through:	Are assigned position or AO by:	Provide liaison to:	Establish/maintain communi-cations with:	Have priorities establish-ed by:	Can impose on gaining unit further command or support relationship of:
Organic	All organic forces organized with the HQ	Organic HQ	Army HQ specified in organizing document	Organic HQ	N/A	N/A	Organic HQ	Attached; OPCON; TACON; GS; GSR; R; DS
Assigned	Combatant command	Gaining HQ	Gaining Army HQ	OPCON chain of command	As required by OPCON	As required by OPCON	ASCC or Service-assigned HQ	As required by OPCON HQ
Attached	Gaining unit	Gaining unit	Gaining Army HQ	Gaining unit	As required by gaining unit	Unit to which attached	Gaining unit	Attached; OPCON; TACON; GS; GSR; R; DS
OPCON	Gaining unit	Parent unit and gaining unit; gaining unit may pass OPCON to lower HQ[1]	Parent unit	Gaining unit	As required by gaining unit	As required by gaining unit and parent unit	Gaining unit	OPCON; TACON; GS; GSR; R; DS
TACON	Gaining unit	Parent unit	Parent unit	Gaining unit	As required by gaining unit	As required by gaining unit and parent unit	Gaining unit	TACON; GS GSR; R; DS

Note: [1] In NATO, the gaining unit may not task-organize a multinational force. (See TACON.)

ADCON	administrative control	HQ	headquarters
AO	area of operations	N/A	not applicable
ASCC	Army Service component command	NATO	North Atlantic Treaty Organization
DS	direct support	OPCON	operational control
GS	general support	R	reinforcing
GSR	general support–reinforcing	TACON	tactical control

B-33. Army commanders normally place a unit OPCON or TACON to a gaining headquarters for a given mission, lasting perhaps a few days. OPCON lets the gaining commander task-organize and direct forces. TACON does not let the gaining commander task-organize the unit. Hence, TACON is the command relationship often used between Army, other Service, and multinational forces within a task organization, but rarely between Army forces. Neither OPCON nor TACON affects ADCON responsibilities. To modify the example used above, if the Army division commander placed the engineer battalion OPCON to the BCT, the gaining brigade commander would not be responsible for the unit training, maintenance, resupply, and unit-level reporting of the engineers. Those responsibilities would remain with the parent maneuver enhancement brigade.

B-34. The ASCC and ARFOR monitor changes in joint organization carefully and may adjust ADCON responsibilities based on the situation. For example, if a joint task force commander places an Army brigade under TACON of a Marine division, the ARFOR may switch some or all unit ADCON responsibilities to another Army headquarters, based on geography and ability to provide administration and support to that Army force.

ARMY SUPPORT RELATIONSHIPS

B-35. Table B-3 lists Army support relationships. Army support relationships are not a command authority and are more specific than the joint support relationships. Commanders establish support relationships when subordination of one unit to another is inappropriate. They assign a support relationship when—

- The support is more effective if a commander with the requisite technical and tactical expertise controls the supporting unit, rather than the supported commander.
- The echelon of the supporting unit is the same as or higher than that of the supported unit. For example, the supporting unit may be a brigade, and the supported unit may be a battalion. It would be inappropriate for the brigade to be subordinated to the battalion, hence the use of an Army support relationship.
- The supporting unit supports several units simultaneously. The requirement to set support priorities to allocate resources to supported units exists. Assigning support relationships is one aspect of mission command.

Table B-3. Army support relationships

If relation-ship is:	Then inherent responsibilities:							
	Have command relation-ship with:	May be task-organized by:	Receives sustain-ment from:	Are assigned position or an area of operations by:	Provide liaison to:	Establish/ maintain communi-cations with:	Have priorities established by:	Can impose on gaining unit further command or support relation-ship by:
Direct support[1]	Parent unit	Parent unit	Parent unit	Supported unit	Supported unit	Parent unit; supported unit	Supported unit	See note[1]
Reinforc-ing	Parent unit	Parent unit	Parent unit	Reinforced unit	Reinforced unit	Parent unit; reinforced unit	Reinforced unit; then parent unit	Not applicable
General support–reinforc-ing	Parent unit	Parent unit	Parent unit	Parent unit	Reinforced unit and as required by parent unit	Reinforced unit and as required by parent unit	Parent unit; then reinforced unit	Not applicable
General support	Parent unit	Parent unit	Parent unit	Parent unit	As required by parent unit	As required by parent unit	Parent unit	Not applicable

Note: [1] Commanders of units in direct support may further assign support relationships between their subordinate units and elements of the supported unit after coordination with the supported commander.

B-36. Army support relationships allow supporting commanders to employ their units' capabilities to achieve results required by supported commanders. Support relationships are graduated from an exclusive supported and supporting relationship between two units—as in direct support—to a broad level of support extended to all units under the control of the higher headquarters—as in general support. Support relationships do not alter ADCON. Commanders specify and change support relationships through task-organizing.

OTHER RELATIONSHIPS

B-37. Several other relationships established by higher headquarters exist with units that are not in command or support relationships. (See table B-4, page B-12.) These relationships are limited or specialized to a greater degree than the command and support relationships. These limited relationships are not used when tailoring or task-organizing Army forces. Use of these specialized relationships helps clarify certain aspects of OPCON or ADCON.

Table B-4. Other relationships

Relation-ship	Operational use	Established by	Authority and limitations
Training and readiness oversight	TRO is an authority exercised by a combatant commander over assigned RC forces not on active duty. Through TRO, CCDRs shape RC training and readiness. Upon mobili-zation of the RC forces, TRO is no longer applicable.	The CCDR identified in the "Forces for Unified Commands" memorandum. The CCDR normally dele-gates TRO to the ASCC. (For most RC forces, the CCDR is JFCOM and the ASCC is FORSCOM.)	TRO allows the CCDR to provide guidance on opera-tional requirements and training priorities, review readiness reports, and review mobilization plans for RC forces. TRO is not a command relationship. ARNG forces remain under the command and control of their respective State Adjutant Generals until mobilized for Federal service. USAR forces remain under the command and control of the USARC until mobilized.
Direct liaison authorized[1]	Allows planning and direct collaboration between two units assigned to different commands, often based on anticipated tailoring and task organization changes.	The parent unit headquarters. This is a coordination relationship, not an authority through which command may be exercised.	Limited to planning and coordination between units.
Aligned	Informal relationship between a theater army and other Army units identified for use in a specific geographic combatant command.	Theater army and parent ASCC.	Normally establishes infor-mation channels between the gaining theater army and Army units that are likely to be committed to that area of responsibility.

Note: [1] See also paragraph B-24.

ARNG	Army National Guard	RC	Reserve Component
ASCC	Army Service component command	TRO	training and readiness oversight
CCDR	combatant commander	USAR	U.S. Army Reserve
FORSCOM	U.S. Army Forces Command	USARC	U.S. Army Reserve Command
JFCOM	Joint Forces Command		

B-38. *Training and readiness oversight* is the authority that combatant commanders may exercise over as-signed Reserve Component forces when not on active duty or when on active duty for training. As a matter of Department of Defense policy, this authority includes: a. Providing guidance to Service component commanders on operational requirements and priorities to be addressed in military department training and readiness programs; b. Commenting on Service component program recommendations and budget re-quests; c. Coordinating and approving participation by assigned Reserve Component forces in joint exer-cises and other joint training when on active duty for training or performing inactive duty for training; d. Obtaining and reviewing readiness and inspection reports on assigned Reserve Component forces; and e. Coordinating and reviewing mobilization plans (including postmobilization training activities and deploy-ability validation procedures) developed for assigned Reserve Component forces (JP 1).

B-39. Responsibilities for both training and readiness are inherent in ADCON and exercised by unit com-manders for their units. Army National Guard forces are organized by the Department of the Army under their respective states. These forces remain under command of the governor of that state until mobilized for Federal service. U.S. Army Reserve forces are assigned to U.S. Army Reserve Command. For Army Na-tional Guard units, the combatant commander normally exercises training and readiness oversight through their ASCC; for most, this is U.S. Army Forces Command. The ASCC coordinates with the appropriate State Adjutants General and Army National Guard divisions to refine mission-essential task lists for Army National Guard units. The ASCC coordinates mission-essential task lists for Army Reserve units with the U.S. Army Reserve Command. When Reserve Component units align with an expeditionary force package during Army force generation, U.S. Army Forces Command establishes coordinating relationships as required between Regular Army and Reserve Component units. When mobilized, Reserve Component units are assigned or attached to their gaining headquarters. Most operating force ADCON responsibilities, including unit training and readiness, shift to the gaining headquarters.

B-40. The shift to full spectrum operations and smaller, more versatile units affects how Regular Army forces manage training and readiness. Army force packages for the combatant commanders combine forces from many different parent organizations through Army force generation. The Army assigns or attaches Regular Army forces to various Army headquarters based on factors such as stationing, unit history, and habitual association of units in training. Different Army headquarters may share ADCON to optimize administration and support. For example, U.S. Army Forces Command may attach a BCT to a division headquarters located on a different installation. That division commander has training and readiness responsibilities for the BCT but does not control the training resources located at the BCT's installation. The senior Army commander on the BCT's installation manages training resources such as ranges and simulation centers. At the direction of the Secretary of the Army, the commanders share ADCON responsibilities. If the division headquarters deploys on an extended mission and the BCT remains, training and readiness responsibilities for the BCT shift to another commander. Headquarters, Department of the Army or another appropriate Army authority redistributes ADCON responsibilities for the BCT to a new headquarters. When the BCT deploys to a geographic combatant command, ADCON passes to the gaining theater army unless modified by the Secretary of the Army. (FMs 7-0 and 7-1 discuss training responsibilities.)

B-41. Alignment is informal relationship between a theater army and other Army units identified for use in the area of responsibility of a specific geographic combatant command. Alignment helps focus unit exercises and other training on a particular region. This may lead to establishment of direct liaison authorized between the aligned unit and a different ASCC. Any modular Army force may find itself included in an expeditionary force package heading to a different combatant command. Therefore, Army commanders maintain a balance between regional focus and global capability.

REGULATORY AUTHORITIES

B-42. Regulations, policies, and other authoritative sources also direct and guide Army forces, Army commands, direct reporting units, ASCCs, and other Army elements. The Army identifies technical matters, such as network operations or contracting, and assigns responsibilities for them to an appropriate organization. These organizations use technical channels established by regulation, policy, or directive. Commanders may also delegate authority for control of certain technical functions to staff officers or subordinate commanders. (FM 6-0 discusses technical channels.)

B-43. The primary regulation governing the missions, functions, and command and staff relationships, including ADCON, of the subordinate elements of the Department of the Army is AR 10-87. This regulation prescribes the relationships and responsibilities among Army forces, Army commands, direct reporting units, and ASCCs. It includes channels for technical supervision, advice, and support for specific functions among various headquarters, agencies, and units. Other regulations and policies specify responsibilities in accordance with Department of Defense directives and U.S. statutes.

This page intentionally left blank.

Appendix C

The Army Modular Force

This appendix provides an overview of Army modular organizations. In 2003, the Army implemented a fundamental shift toward a brigade-based force. The ongoing transformation of the Army will result in stand-alone division and corps headquarters. Brigade combat teams, modular support brigades, and functional brigades will be pooled for use as part of expeditionary force packages that enhance the flexibility and responsiveness of the Army. The combined arms brigade combat teams become the centerpiece for Army maneuver. They will attach to a higher echelon headquarters—a division, corps, or theater army—as part of a force-tailored formation based on operational requirements.

BACKGROUND

C-1. Today's operational environment requires responsive Army forces tailored to individual combatant commanders' needs. The highly integrated organization of the Army's divisions in the late 1990's made it difficult to deploy divisional units apart from their divisional base and keep the rest of the division ready for other missions. Coupled with the increasing need to employ land forces at the outset of a campaign, the Army needed to reorganize around smaller, more versatile formations able to deploy more promptly.

C-2. No single, large fixed formation can support the diverse requirements of full spectrum operations. To meet the requirements of the geographic combatant commanders, the Army has developed the capability to rapidly tailor and task-organize expeditionary force packages. A force package may consist of any combination of light, medium, and heavy forces; it can blend Regular Army, Army National Guard, and U.S. Army Reserve units and Soldiers.

C-3. The nature of modern land operations has changed in geography and time. In general, operations have become increasingly distributed in space while more simultaneous in time. At the tactical and operational levels, subordinate units routinely operate in noncontiguous areas of operations. This contrasts sharply with the contiguous and hierarchical arrangement of land forces in operations prevalent in the past. More agile forces, improvements in command and control, and continuing integration of joint capabilities at lower echelons all contribute to these changes.

C-4. The other prominent shift in capability came with the introduction and proliferation of satellite-based communications and other advanced information systems for command and control. Command and control of widely dispersed formations no longer entirely relies on terrestrial, line-of-sight communications. When separated by hundreds of miles, today's commanders can still communicate with subordinates and maintain a common operational picture. The Army is only beginning to realize the benefits of these advances. It continues to leverage technology and reshape processes to best integrate new capabilities.

C-5. Tactical operations continue to evolve into distributed, noncontiguous forms. Army forces need versatile and deployable headquarters suited for contingencies and protracted operations. The Army provides the majority of land component command headquarters and joint task force headquarters for contingency operations. The complexity of counterinsurgency campaigns, such as those in Afghanistan and Iraq, require Army headquarters to function as joint and multinational platforms. While dealing with complex issues, the headquarters deploy, evolve, and tailor their compositions as the campaign progresses. As recent natural disasters showed, Army headquarters often provide the command and control element for Regular Army, Army National Guard, and U.S. Army Reserve elements that respond to disasters of all types.

C-6. To meet joint requirements, the Army reorganized its operating forces beginning in 2003. Today, the Army can provide land combat power tailored for any combination of offensive, defensive, and stability or civil support operations as part of an interdependent joint force. Brigades are the principal tactical units for conducting operations. To provide higher echelon command and control, the Army fields a mix of tactical and operational headquarters able to function as land force, joint, multinational, and Service component command headquarters. The headquarters mix is not a rigid hierarchy and does not require a standard array of forces. Each headquarters provides a menu of capabilities to best match the combatant commander's requirements.

C-7. The combatant commanders' requirements are determined by the National Military Strategy, the Joint Strategic Capabilities Plan (as specified in the "Forces for" portions), and operational requirements. The strategic Army role of providing forces to meet global requirements is called force generation. As part of force generation, the Department of the Army establishes manning, training, and readiness cycles; assigns forces to headquarters; and manages modernization. Strategic organization establishes goals for force generation cycles based on Regular Army, Army National Guard, and U.S. Army Reserve manning and readiness cycles.

DIVISION ECHELON AND ABOVE

C-8. The Army of Excellence structure for headquarters and large formations has evolved into three modular headquarters organizations. The Army Service component command focuses on combatant command-level landpower employment. It supports joint, interagency, intergovernmental, and multinational forces within a combatant commander's area of responsibility. The corps provides a headquarters that specializes in operations as a land component command headquarters, as a joint task force for contingencies, or as an intermediate tactical headquarters within large groupings of land forces. The division is optimized for tactical control of brigades during land operations. All three headquarters are modular entities designed to use forces tailored for specified joint operations. All three are also stand-alone headquarters unconstrained by a fixed formation of subordinate units. While three types of modular headquarters exist, the Army forces they control are organized for two broad echelons—theater and tactical. Each set consists primarily of brigades.

THEATER ARMY HEADQUARTERS COMMANDS

C-9. The doctrinal name for the Army Service component command of a geographic combatant command is theater army. The theater army is the primary vehicle for Army support to Army, joint, interagency, intergovernmental, and multinational forces operating across the area of responsibility. When the combatant commander acts as the joint force commander during major combat operations, the theater army may provide the land component commander and headquarters. In that case, it exercises operational control (OPCON) over land forces deployed to a joint operations area. The theater army headquarters continues to perform area of responsibility-wide functions in addition to its operational responsibilities. These functions include reception, staging, onward movement, and integration; logistics over-the-shore operations; and security coordination. (Figure C-1 shows an example of a theater army headquarters organized as a land component command.) When required, the theater army can provide a headquarters able to command and control a joint task force for contingencies with other Service augmentation.

C-10. As the Army Service component command, the theater army exercises administrative control (ADCON) over all Army forces in the area of responsibility unless modified by the Department of the Army. This includes forces assigned, attached, or OPCON to the combatant command. The Army Service component command provides Army support to designated theater-level forces. It also provides this support to joint, interagency, intergovernmental, and multinational elements as the combatant commander directs. The Army Service component command integrates Army forces into execution of theater security cooperation plans as well. It has several theater-level formations associated with it.

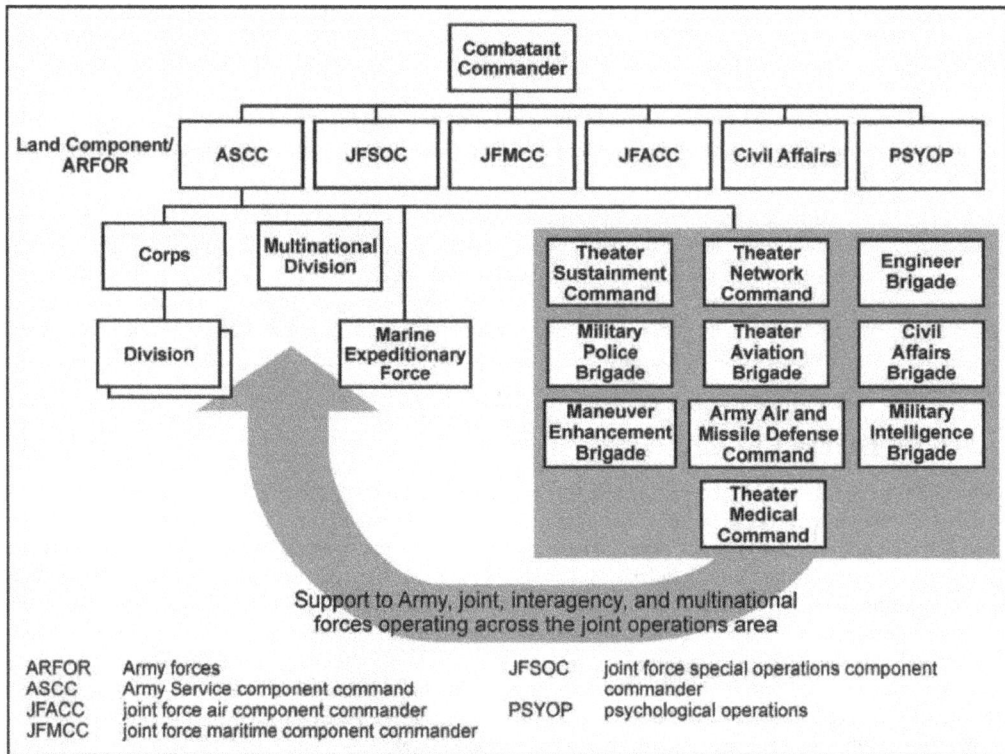

Figure C-1. Example of theater army acting as a land component command while continuing Army support

THEATER-LEVEL FORMATIONS

C-11. An array of theater-level forces may be assigned, attached, or OPCON to a theater army headquarters. Each theater army headquarters normally has organizations providing theater-level capabilities aligned with it or under its control. These organizations include—

- Theater sustainment command.
- Military intelligence brigade.
- Theater network command or brigade.
- Regionally focused civil affairs brigade or planning team.
- Regionally focused medical command.
- Regionally focused air and missile defense command.

C-12. When the theater army is the land component command for major combat operations, several functional commands may augment it. These commands can include—

- Engineer.
- Military police.
- Criminal investigation.
- Aviation.

These commands consist of units from the Regular Army, Army National Guard, and U.S. Army Reserve.

C-13. Several functional brigades are also available to support theater-level operations. They may be task-organized under theater-level functional commands or be directly subordinate to the theater army. When required, the theater army may task-organize functional brigades to corps or divisions. Examples of functional brigades include the following:

- Civil affairs.
- Engineer.
- Theater aviation.
- Military police.
- Chemical, biological, radiological, and nuclear.
- Air and missile defense.
- Medical.

CORPS

C-14. Large land forces require an intermediate echelon between the divisions that control brigade combat teams (BCTs) and the theater army serving as the land component command. Other factors requiring an intermediate headquarters may include—

- The mission's complexity.
- Multinational participation.
- Span of control.

C-15. When required, a corps may become an intermediate tactical headquarters under the land component command, with OPCON of multiple divisions (including multinational or Marine Corps formations) or other large tactical formations. (See figure C-2.) The theater army headquarters tailors the corps headquarters to meet mission requirements. The corps' flexibility allows the Army to meet the needs of joint force commanders for an intermediate land command while maintaining a set of headquarters for contingencies.

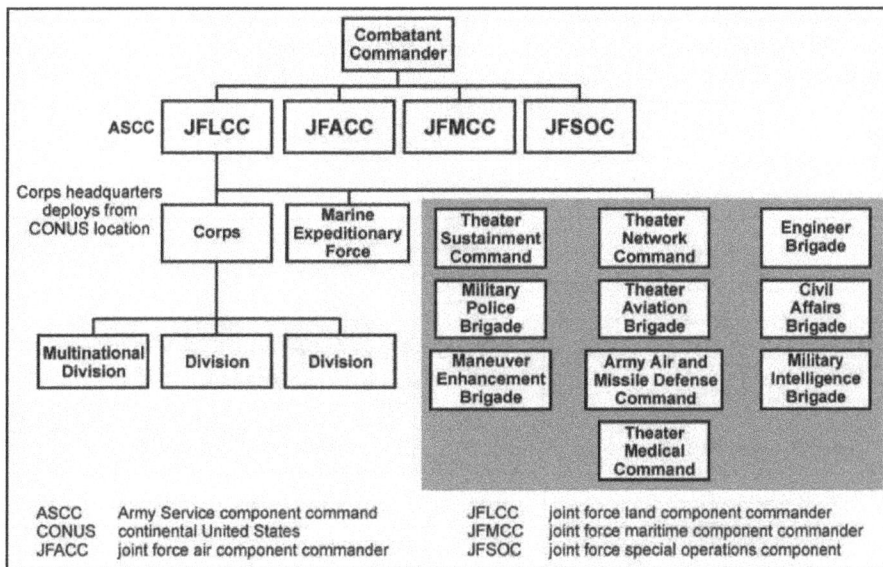

Figure C-2. Corps as an intermediate land force headquarters

C-16. The corps is also a primary candidate headquarters for joint operations. It can rapidly transition to either a joint task force or land component command headquarters for contingency or protracted operations. It can deploy to any area of responsibility to provide command and control for Army, joint, and multinational forces. The corps does not have any echelon-specific units other than the organic corps headquarters. It can control any mix of modular brigades and divisions, as well as other Service or multinational forces. When used as a land component or as an intermediate tactical headquarters, the corps may also be designated as the ARFOR, with ADCON responsibility for all Army forces subordinate to the joint task force. (Appendix B discusses ARFOR and ADCON responsibilities.)

C-17. When directed, a corps trains as a joint headquarters for contingency operations. With minimum joint augmentation, this headquarters can initiate operations as a joint task force or land component command for contingencies. For sustained operations in this role, a corps is augmented according to an appropriate joint manning document. The corps can also serve as a deployable base for a multinational headquarters directing protracted operations.

DIVISION

C-18. Divisions are the Army's primary tactical warfighting headquarters. Their principal task is directing subordinate brigade operations. Divisions are not fixed formations. They exercise command and control over any mix of brigades and do not have any organic forces beyond their headquarters elements. Their organic structure includes communications network, life support, and command post elements. These provide significant flexibility. With appropriate joint augmentation, a division can be the joint task force or land component command headquarters for small contingencies. The headquarters staff has a functional organization. It also includes organic joint network capability and liaison teams.

C-19. Divisions can control up to six BCTs in major combat operations. They can control more BCTs in protracted stability operations. A division force package may include any mix of heavy, infantry, and Stryker BCTs. In addition to BCTs, each division controls a tailored array of modular support brigades and functional brigades.

C-20. Division headquarters normally have four BCTs attached for training and readiness purposes. However, these brigades may or may not deploy with the division as part of an expeditionary force package. Since divisions have no organic structure beyond the headquarters, all types of brigades may not be present in an operation. In some operations, divisions may control multiple support brigades of the same type. They may also control functional groups, battalions, or separate companies; however, these are normally task-organized to a brigade. The important point is that division organizations vary for each operation. However, for major combat operations, divisions should have at least one of each type of support brigade attached or OPCON to it. (Figures C-3 and C-4, page C-6, illustrate two possible division organizations. Many more combinations are possible.)

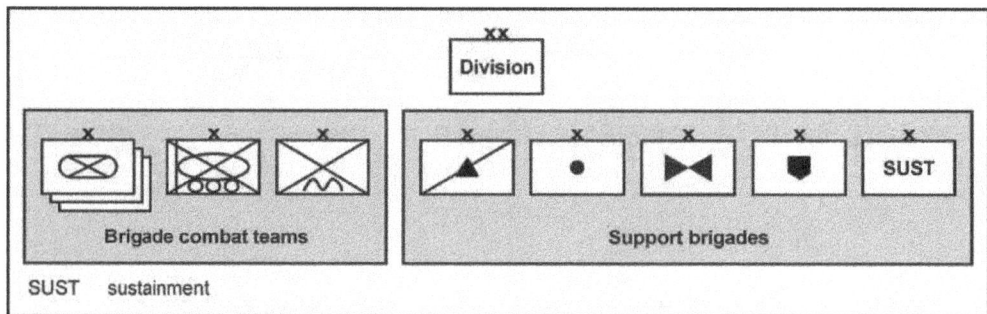

Figure C-3. Example of tailored divisions in offensive operations

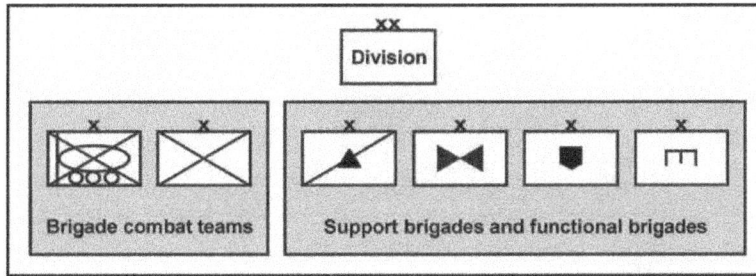

Figure C-4. Example of tailored divisions in defensive operations

BRIGADE COMBAT TEAMS

C-21. As combined arms organizations, BCTs form the basic building block of the Army's tactical formations. They are the principal means of executing engagements. Three standardized BCT designs exist: heavy, infantry, and Stryker. Battalion-sized maneuver, fires, reconnaissance, and sustainment units are organic to BCTs.

C-22. BCTs are modular organizations. They begin as a cohesive combined arms team that can be further task-organized. Commands often augment them for a specific mission with capabilities not organic to the BCT structure. Augmentation might include lift or attack aviation, armor, cannon or rocket artillery, air defense, military police, civil affairs, psychological operations elements, combat engineers, or additional information systems assets. This organizational flexibility allows BCTs to function across the spectrum of conflict.

C-23. The Army plans to convert BCTs to very advanced combined arms formations equipped with the family of future combat systems. These highly modernized brigades will consist of three combined arms battalions, a non-line-of-sight cannon battalion, reconnaissance surveillance and target acquisition squadron, brigade support battalion, brigade intelligence and communications company, and a headquarters company. The brigade combat teams equipped with future combat systems will improve the strategic and operational reach of ground combat formations without sacrificing lethality or survivability. Well before the future combat systems brigades join the operating forces, the Army will field some advanced systems to the current force.

HEAVY BRIGADE COMBAT TEAM

C-24. Heavy BCTs are balanced combined arms units that execute operations with shock and speed. (See figure C-5.) Their main battle tanks, self-propelled artillery, and fighting vehicle-mounted infantry provide tremendous striking power. Heavy BCTs require significant strategic air- and sealift to deploy and sustain. Their fuel consumption may limit operational reach. However, this is offset by the heavy BCT's unmatched tactical mobility and firepower. Heavy BCTs include organic military intelligence, artillery, signal, engineer, reconnaissance, and sustainment capabilities.

INFANTRY BRIGADE COMBAT TEAM

C-25. The infantry BCT requires less strategic lift than other BCTs. (See figure C-6.) When supported with intratheater airlift, infantry BCTs have theaterwide operational reach. The infantry Soldier is the centerpiece of the infantry BCT. Organic antitank, military intelligence, artillery, signal, engineer, reconnaissance, and sustainment elements allow the infantry BCT commander to employ the force in combined arms formations. Infantry BCTs work best for operations in close terrain and densely populated areas. They are easier to sustain than the other BCTs. Selected infantry BCTs include special-purpose capabilities for airborne or air assault operations.

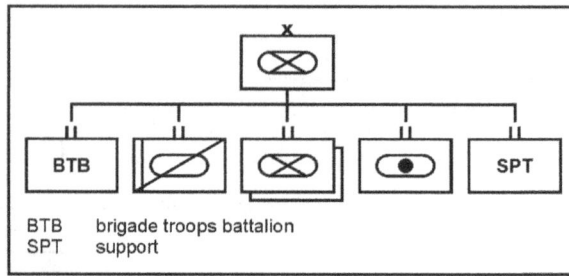

Figure C-5. Heavy brigade combat team

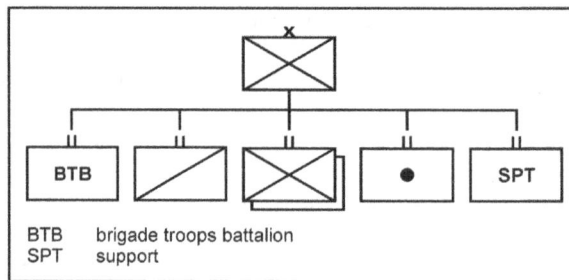

Figure C-6. Infantry brigade combat team

STRYKER BRIGADE COMBAT TEAM

C-26. The Stryker BCT balances combined arms capabilities with significant strategic and intratheater mobility. (See figure C-7.) Designed around the Stryker wheeled armored combat system in several variants, the Stryker BCT has considerable operational reach. It is more deployable than the heavy BCT and has greater tactical mobility, protection, and firepower than the infantry BCT. Stryker BCTs have excellent dismounted capability. The Stryker BCT includes military intelligence, signal, engineer, antitank, artillery, reconnaissance, and sustainment elements. This design lets Stryker BCTs commit combined arms elements down to company level in urban and other complex terrain against a wide range of opponents.

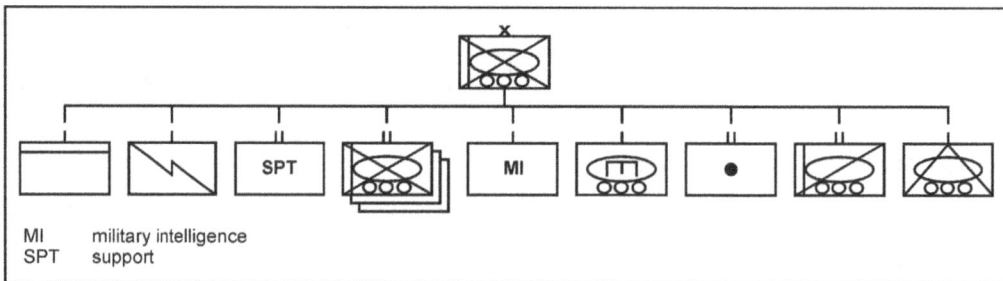

Figure C-7. Stryker brigade combat team

MODULAR SUPPORT BRIGADES

C-27. Five types of modular support brigades complement the BCTs: the battlefield surveillance brigade, fires brigade, combat aviation brigade, maneuver enhancement brigade, and sustainment brigade. These brigades provide multifunctional capabilities to deployed forces. More than one type of support brigade may be task-organized to a division or corps (except for sustainment brigades, which provide general or direct support to a division or corps). In turn, commands may make these brigades available to other Service components of the joint force. Support brigades have the organic expertise to command and control various unit types. Theater armies tailor them by adding functional battalions to or subtracting them from the organic command and control headquarters. The signal and maintenance capabilities of a support brigade headquarters also allows the higher headquarters to task-organize them to a corps, headquarters of another Service, or joint headquarters.

C-28. The number and type of subordinate units vary among the different types of brigades. Four types of support brigades operate as part of a division-sized expeditionary force: the battlefield surveillance brigade, fires brigade, combat aviation brigade, and maneuver enhancement brigade. They are normally assigned, attached, or placed OPCON to a division. Normally, the theater army attaches the sustainment brigade to the theater sustainment command. This brigade provides either general or direct support to forces under the divisions. For a major combat operation, the higher headquarters normally task-organizes one of each type of the five support brigades to a division headquarters.

BATTLEFIELD SURVEILLANCE BRIGADE

C-29. The battlefield surveillance brigade has military intelligence, reconnaissance and surveillance, and requisite sustainment and communications capabilities. (See figure C-8.) The headquarters and headquarters company provides command and control of brigade operations. The military intelligence battalion provides unmanned aircraft systems, signals intelligence, human intelligence, and counterintelligence capabilities. The reconnaissance and surveillance battalion provides reconnaissance and surveillance capabilities, including mounted scout platoons and mobile long-range surveillance teams. The brigade support company provides sustainment for the brigade. The network company provides a communications backbone. This allows the battlefield surveillance brigade to communicate throughout the division area of operations as well as with support assets associated with Army Service component command- and national-level intelligence agencies. Battlefield surveillance brigades can be tailored for the mission before deployment or task-organized by the higher headquarters once deployed. Typical augmentation includes—

- Ground reconnaissance.
- Manned and unmanned Army aviation assets.
- Military intelligence assets, including human intelligence, aerial exploitation, and other national-level assets.
- Armored, infantry, and combined arms units.

C-30. The battlefield surveillance brigade conducts intelligence, surveillance, and reconnaissance (ISR) operations. This capability lets the division commander focus combat power, execute current operations, and prepare for future operations simultaneously. Battlefield surveillance brigades are not designed to conduct guard or cover operations. Those operations may entail fighting to develop the tactical situation; they require a BCT or aviation brigade.

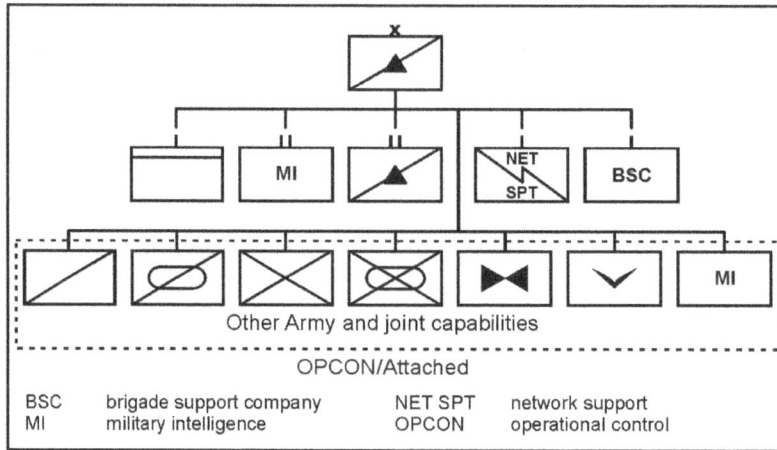

Figure C-8. Battlefield surveillance brigade

FIRES BRIGADE

C-31. Fires brigades are normally assigned, attached, or OPCON to a division. However, they may be OPCON to a task force, land component command, or other Service or functional component. (See figure C-9.) Fires brigades are task-organized based on their assigned tasks.

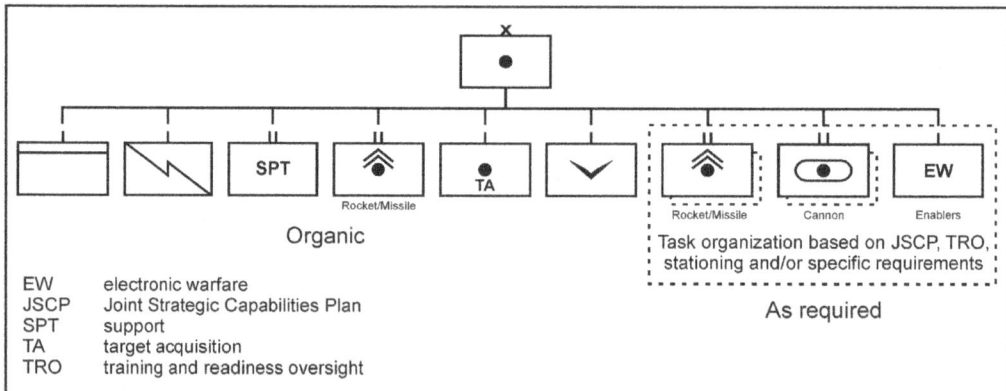

Figure C-9. Fires brigade

C-32. Organic fires brigade assets include a Multiple Launch Rocket System battalion, headquarters battery, and target acquisition battery. Brigades may be task-organized with additional Multiple Launch Rocket System and cannon battalions, counterfire radars, and joint information operations assets. The brigade's higher headquarters usually assigns the brigade missions in terms of target sets to engage, target priorities, or effects to achieve. The situation may also require the brigade to control joint fires assets.

C-33. A fires brigade's primary task is conducting strike operations. This task requires placing ISR and electronic attack capabilities OPCON to the brigade headquarters. Alternatively, the battlefield surveillance brigade can retain control of ISR assets and provide targeting information to the fires brigade through a support relationship.

C-34. Fires brigades perform the following tasks:

- Conduct strike operations.
- Support BCTs and other brigades.
- Conduct joint missions separate from the division.
- Conduct fire support missions for the division and brigades, including counterfire and attacks on specific targets in the division's area of operations.

COMBAT AVIATION BRIGADE

C-35. Most of the Army's aviation combat power resides in multifunctional combat aviation brigades. These organizations can be task-organized based on the mission. They include various types of organizations, with manned and unmanned systems. Combat aviation brigades are organized to support divisions, BCTs, and support brigades. (See figure C-10.) They specialize in providing combat capabilities to multiple BCTs. However, they can be task-organized to support a theater army or corps acting as a joint task force or land component command.

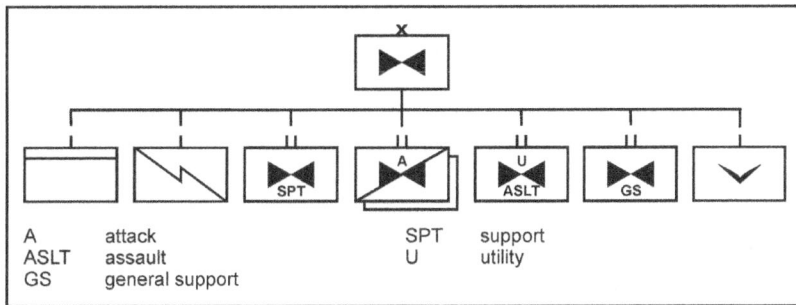

Figure C-10. Combat aviation brigade

C-36. Based on priorities and missions, a combat aviation brigade coordinates operational details directly with supported organizations. Combat aviation brigades typically conduct the following missions:

- Attack.
- Reconnaissance.
- Security.
- Movement to contact.
- Air assault.
- Air movement.
- Aerial casualty evacuation.
- Personnel recovery.
- Command and control support.

SUSTAINMENT BRIGADE

C-37. Sustainment brigades normally have a command relationship with a theater sustainment command and provide general or direct support to divisions and brigades. In major combat operations, the sustainment brigade may be under OPCON of or provide direct support to a division. Sustainment brigades have a flexible organization designed to be task-organized to meet mission requirements. (See figure C-11.) They have a command and staff structure able to control operational- or tactical-level sustainment.

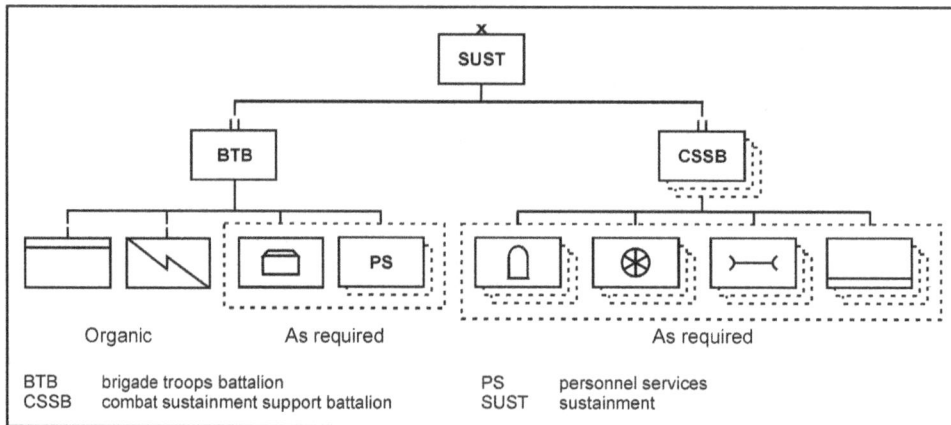

Figure C-11. Sustainment brigade

C-38. The higher headquarters usually reinforces the sustainment brigade with several different modular sustainment elements. The types and quantities of these attachments depend on the mission and the number, size, and type of organizations requiring support.

C-39. A sustainment brigade's only organic unit is its brigade troops battalion. This battalion provides command and control for assigned and attached personnel and units. It directs sustainment operations for the brigade headquarters.

MANEUVER ENHANCEMENT BRIGADE

C-40. Maneuver enhancement brigades command and control forces that provide protection and other support to the force. They are tailored with the capabilities required for each operation. More than one brigade may be assigned to a division or corps. (See figure C-12.) Commands may also attach these brigades directly to the theater army to serve in the theater army area of operations or joint security area.

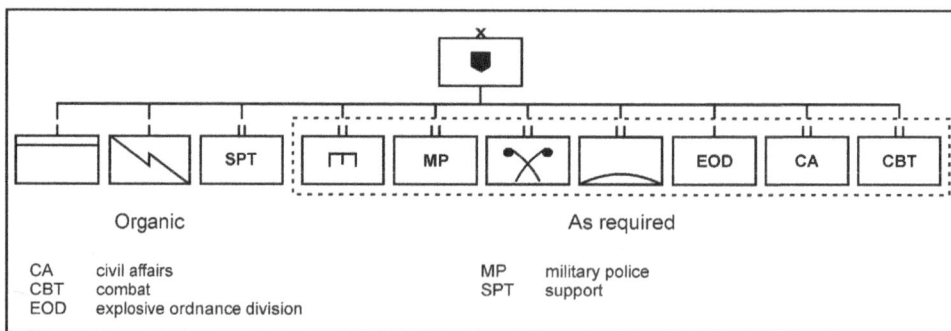

Figure C-12. Maneuver enhancement brigade

C-41. Maneuver enhancement brigades are designed to control the following types of organizations:
- Engineer.
- Military police.
- Chemical, biological, radiological, and nuclear.
- Civil affairs.

- Air and missile defense.
- Explosive ordnance disposal.
- Tactical combat force (when given an area security mission).

C-42. Typical missions for maneuver enhancement brigades are to—
- Conduct area security operations.
- Construct, maintain, and sustain lines of communications.
- Provide mobility and countermobility support.
- Provide vertical, runway, and road construction.
- Conduct chemical, biological, radiological, and nuclear defense throughout the area of operations.
- Conduct limited offensive and defensive tasks.
- Conduct some stability tasks.
- Conduct consequence management operations.

C-43. Maneuver enhancement brigades are organized and trained to conduct selected area security missions, including route and convoy security operations. They are not designed to screen, guard, or cover. A maneuver enhancement brigade may be assigned a support area encompassing the supporting and sustainment organizations and main supply routes of the supported command, typically a division. This mission does not supplant unit local security responsibilities. Units remain responsible for self-protection against level I threats. (JP 3-10 and FM 3-90 discuss area security and protection, including threat levels and tactical combat forces. FMI 3-0.1 discusses support areas.)

C-44. Maneuver enhancement brigades can employ a maneuver battalion as a tactical combat force when the situation requires. With a tactical combat force, the brigade executes limited offensive and defensive tasks against level II or III threats. Tactical combat forces may include not only ground maneuver but also aviation and fires assets. However, commanders should employ a BCT when the situation requires a tactical combat force of two or more ground maneuver battalions.

FUNCTIONAL BRIGADES

C-45. Functional brigades, like the modular support brigades, have a modular subordinate structure that may vary considerably among brigades of the same type. Unlike the modular support brigades, functional brigades typically operate under theater army control and depend on theater-level elements for signal and other support. The theater army may task-organize them to corps or division headquarters. Types of functional brigades include—
- Engineer.
- Military police.
- Chemical, biological, radiological, and nuclear.
- Air and missile defense.
- Signal.
- Explosive ordnance disposal.
- Medical.
- Intelligence.

MODULAR ARMY FORCES CONTROLLED BY OTHER SERVICES

C-46. Other Service headquarters may control modular Army brigades directly. A division or corps headquarters is not necessary. Figure C-13 illustrates a maneuver enhancement brigade OPCON to a Marine expeditionary force. The theater army, with its assigned commands, continues to exercise ADCON over the brigade. The theater army also provides Army capabilities, such as network operations, in support of the Marine expeditionary force. Other examples include placing fires brigade equipped with a Multiple Launch

Rocket System OPCON to the joint force air component commander or a tailored sustainment brigade in general or direct support to a joint special operations task force.

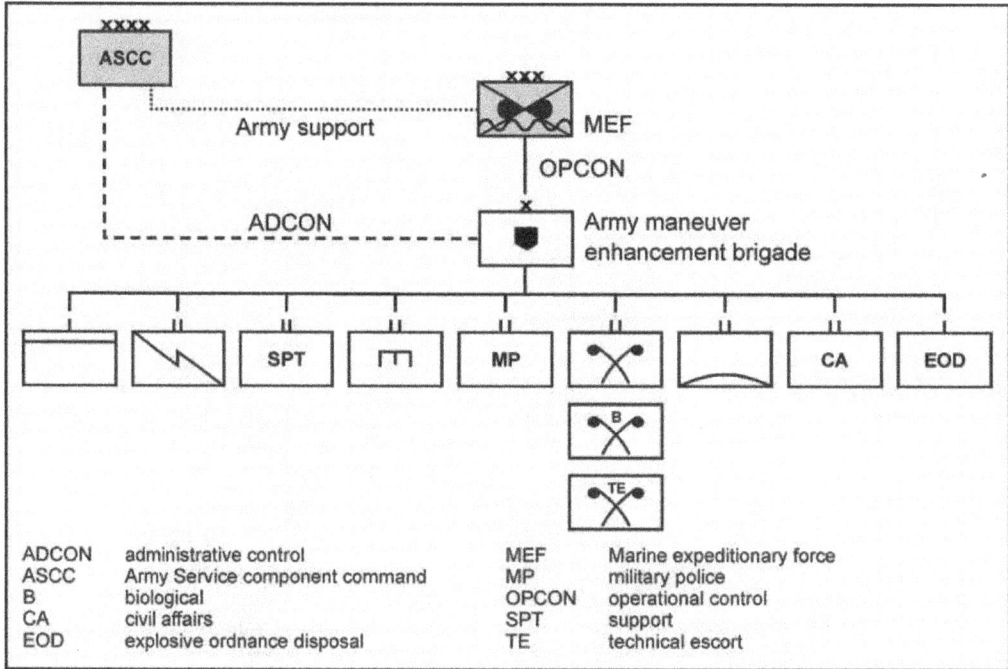

Figure C-13. Maneuver enhancement brigade OPCON to a Marine expeditionary force

This page intentionally left blank.

Appendix D

The Role of Doctrine and Summary of Changes

This appendix discusses the role of doctrine in full spectrum operations and describes the major doctrinal changes contained in this manual.

THE ROLE OF DOCTRINE

D-1. Army doctrine is a body of thought on how Army forces intend to operate as an integral part of a joint force. Doctrine focuses on how to think—not what to think. It establishes the following:

- How the Army views the nature of operations.
- Fundamentals by which Army forces conduct operations.
- Methods by which commanders exercise command and control.

D-2. Doctrine is a guide to action, not a set of fixed rules. It combines history, an understanding of the operational environment, and assumptions about future conditions to help leaders think about how best to accomplish missions. Doctrine is consistent with human nature and broad enough to provide a guide for unexpected situations. It is also based upon the values and ethics of the Service and the Nation; it is codified by law and regulations and applied in the context of operations in the field. It provides an authoritative guide for leaders and Soldiers but requires original applications that adapt it to circumstances. Doctrine should foster initiative and creative thinking.

D-3. Doctrine establishes a common frame of reference including intellectual tools that Army leaders use to solve military problems. It is a menu of practical options based on experience. By establishing common approaches to military tasks, doctrine promotes mutual understanding and enhances effectiveness. It facilitates communication among Soldiers and contributes to a shared professional culture. By establishing a commonly understood set of terms and symbols, doctrine facilitates rapid dissemination of orders and fosters collaborative synchronization among units. It establishes the foundation for curricula in the Army Education System.

D-4. Army doctrine forms the basis for training and leader development standards and support products. Training standards provide performance baselines to evaluate how well a task is executed. Together, doctrine, training, and resources form the key to Army readiness. Doctrine consists of—

- Fundamental principles.
- Tactics, techniques, and procedures.
- Terms and symbols.

FUNDAMENTAL PRINCIPLES

D-5. Fundamental principles provide the foundation upon which Army forces guide their actions. They foster the initiative needed for leaders to become adaptive, creative problem solvers. These principles reflect the Army's collective wisdom regarding past, present, and future operations. They provide a basis for incorporating new ideas, technologies, and organizational designs. Principles apply at all levels of war.

TACTICS, TECHNIQUES, AND PROCEDURES

D-6. Principles alone are not enough to guide operations. Tactics, techniques, and procedures provide additional detail and more specific guidance, based on evolving knowledge and experience. Tactics, techniques, and procedures support and implement fundamental principles, linking them with associated applications. The "how to" of tactics, techniques, and procedures includes descriptive and prescriptive methods and processes. Tactics, techniques, and procedures apply at the operational and tactical levels.

D-7. *Tactics* is the employment and ordered arrangement of forces in relation to each other (CJCSI 5120.02A). Effective tactics translate combat power into decisive results. Primarily descriptive, tactics vary with terrain and other circumstances; they change frequently as the enemy reacts and friendly forces explore new approaches. Applying tactics usually entails acting under time constraints with incomplete information. Tactics always require judgment. For example, a commander may choose to suppress an enemy with fires delivered by the majority of the force while maneuvering a small element to envelop the enemy's position. In a general sense, tactics concerns the application of the various primary tasks associated with those elements of full spectrum operations discussed in FM 3-0. (When revised, FM 3-07 will detail tactical tasks associated with stability operations. A field manual addressing the primary civil support tasks is under development. FM 3-90 will address the primary tasks for offensive and defensive operations.)

D-8. Employing a tactic usually requires using and integrating several techniques and procedures. *Techniques* are nonprescriptive ways or methods used to perform missions, functions, or tasks (CJCSI 5120.02A). They are the primary means of conveying the lessons learned that units gain in operations. Commanders base the decision to use a given technique on their assessment of the situation.

D-9. *Procedures* are standard, detailed steps that prescribe how to perform specific tasks (CJCSI 5120.02A). They normally consist of a series of steps in a set order. Procedures are prescriptive; regardless of circumstances, they are executed in the same manner. Techniques and procedures are the lowest level of doctrine. They are often based on equipment and are specific to particular types of units.

TERMS AND SYMBOLS

D-10. Doctrine provides a common language for professionals to communicate with one another. Terms with commonly understood definitions comprise a major part of that language. Symbols are the language's graphic representations. Establishing and using words and symbols with common military meanings enhances communication among professionals. It makes a common understanding of doctrine possible. Definitions for military terms are established in joint publications, field manuals, and field manuals-interim. The field manual or field manual-interim that establishes an Army term's definition is the proponent publication for that term. Terms are listed in JP 1-02 and FM 1-02. FM 1-02 is also the proponent field manual for symbols. Symbols are always prescriptive. Effective command and control requires terms and symbols that are commonly understood, regardless of Service.

EFFECTS AND ARMY DOCTRINE

D-11. Army forces conduct operations according to Army doctrine. The methods that joint force headquarters use to analyze an operational environment, develop plans, or assess operations do not change this. During operations, joint force headquarters provide direction to senior Army headquarters. Army headquarters then perform the military decisonmaking process (MDMP) to develop its own plan or order. (FM 5-0 describes the MDMP.)

D-12. Army forces do not use the joint systems analysis of the operational environment, effects-based approach to planning, or effects assessment. These planning and assessment methods are intended for use at the strategic and operational levels by properly resourced joint staffs. However, joint interdependence requires Army leaders and staffs to understand joint doctrine that addresses these methods when participating in joint operation planning or assessment or commanding joint forces. (JPs 3-0 and 5-0 establish this doctrine.)

D-13. Describing and assessing operations in terms of effects does not fundamentally change Army doctrine. Army operations remain purpose based and conditions focused. The fundamentals of full spectrum operations and mission command include the idea of focusing efforts toward establishing conditions that define the end state. Achieving success in operations requires commanders to gauge their progress continually. Assessing whether tasks are properly executed cannot accomplish this alone. Rather, commanders assess an operation's progress by evaluating how well the results of executing various tasks contribute to creating end state conditions.

SUMMARY OF MAJOR CHANGES

D-14. The following paragraphs summarize the major doctrinal changes made by this field manual.

CHAPTER 1 – THE OPERATIONAL ENVIRONMENT

D-15. Chapter 1 makes the following changes:

- Replaces the dimensions of the operational environment with the variables established in JP 3-0 (political, military, economic, social, information, infrastructure) plus physical environment and time (**PMESII-PT**). Together, these make up the *operational variables*. The factors of a specific situation bounded by assignment of a mission remain mission, enemy, terrain and weather, troops and support available, time available, civil considerations (**METT-TC**). Together, these are called the *mission variables*.
- Lists areas of **joint interdependence**.
- Incorporates the **Soldier's Rules** established in AR 350-1.

CHAPTER 2 – THE CONTINUUM OF OPERATIONS

D-16. Chapter 2 makes the following changes:

- Establishes the **spectrum of conflict** as a way to describe the level of violence in the operational environment.
- Establishes **operational themes** as a means to describe the character of the predominant major operation within a land force commander's area of operations. The operational themes also provide a framework for categorizing the various types of operations described in joint doctrine.

CHAPTER 3 – FULL SPECTRUM OPERATIONS

D-17. Chapter 3 makes the following changes:

- States the Army's **operational concept** and describes its place in doctrine.
- Describes the Army's role in homeland security.
- Changes the approach to **stability operations**. Stability operations are considered coequal with offensive and defensive operations. They are now discussed in terms of five tactical tasks.
- Rescinds **support operations** as a type of operation. Establishes **civil support operations** as an element of full spectrum operations conducted only in the United States and its territories.
- Uses **lethal** and **nonlethal** as broad descriptions of actions. Rescinds the terms *kinetic* and *nonkinetic*.

CHAPTER 4 – COMBAT POWER

D-18. Chapter 4 makes the following changes:

- Replaces the **battlefield operating systems** with the **warfighting functions** (movement and maneuver, intelligence, fires, sustainment, command and control, and protection).
- Retains the fundamental of **combat power** but changes the **elements of combat power** to the six warfighting functions tied together by leadership and information.
- Rescinds the terms *combat arms, combat support,* and *combat service support.* Uses the appropriate warfighting function to describe unit types and functions.
- Rescinds the *tenets of operations.* The warfighting functions and elements of combat power perform the function of this fundamental.

CHAPTER 5 – COMMAND AND CONTROL

D-19. Chapter 5 makes the following changes:

- Modifies the definition of **command** for the Army. The definition now includes leadership.
- Prescribes a new definition of **battle command**.
- Prescribes a new definition of **commander's visualization**.
- Prescribes a new definition of **commander's intent**.
- Adds *understand* to the commander's role in battle command (described in FM 3-0 [2001] as visualize, describe, direct, assess, and lead).
- Rescinds the *operational framework* construct, including its subordinate constructs of *battlespace* and *battlefield organization*. (**Area of operations** is retained.) Retains **decisive, shaping,** and **sustaining operations** (formerly the purpose-based *battlefield organization*) and **main effort** as ways commanders describe subordinates' actions in the concept of operations.
- Prescribes the term **unassigned area** to designate areas between noncontiguous areas of operations or beyond contiguous areas of operations. The higher headquarters is responsible for controlling unassigned areas in its area of operations.
- Rescinds the terms *deep, close,* and *rear areas*. Uses **close combat** to describe operations in what used to be called the close area.
- Eliminates *linear* and *nonlinear* as ways to describe the array of forces on the ground. Army doctrine now describes force arrays as occupying either contiguous or noncontiguous areas of operations.
- Describes how the operations process includes several **integrating processes** and **continuing activities** that commanders and staffs synchronize throughout operations.
- Replaces the term *criteria of success* with the joint terms **measure of effectiveness** and **measure of performance**.

CHAPTER 6 – OPERATIONAL ART

D-20. Chapter 6 makes the following changes:

- Introduces **problem framing** as fundamental to operational art.
- Incorporates **risk** as an element of operational design.
- Prescribes the terms **defeat mechanism** and **stability mechanism**. Establishes individual defeat and stability mechanisms.
- Prescribes the term **line of effort** to replace the term *logical line of operations*.

CHAPTER 7 – INFORMATION SUPERIORITY

D-21. Chapter 7 makes the following changes:

- Adds **knowledge management** as a contributor to information superiority.
- Prescribes the following terms:
  - **Intelligence, surveillance, and reconnaissance (ISR).**
  - **ISR integration.**
  - **ISR synchronization.**
  - **Command and control warfare.**
  - **Information engagement.**
- Describes how Army forces uses five information tasks to shape the operational environment:
  - Information engagement.
  - Command and control warfare.
  - Information protection.
  - Operations security.

- Military deception.
- Adopts the term **situational awareness**.

CHAPTER 8 – STRATEGIC AND OPERATIONAL REACH

D-22. Chapter 8 describes the strategic and operational reach (formerly known as strategic responsiveness) that have been developed since FM 3-0 (2001) was published.

APPENDIX A – PRINCIPLES OF WAR AND OPERATIONS

D-23. Appendix A adds the following joint principles of operations to the principles of war: perseverance, legitimacy, and restraint.

APPENDIX B – COMMAND AND SUPPORT RELATIONSHIPS

D-24. Appendix B discusses administrative control and how it applies to tailored and task-organized Army forces. It also explains how headquarters share administrative control.

APPENDIX C – THE ARMY MODULAR FORCE

D-25. Appendix C describes the modular organizations developed since FM 3-0 (2001) was published.

TERMS AND DEFINITIONS

D-26. The following tables list changes to terms for which FM 3-0 is the proponent field manual. Army terms that also have a joint definition are followed by *(Army)*. Terms for which the Army and Marine Corps have agreed on a common definition are followed by *(Army/Marine Corps)*. The tables do not show changed terms when the changes are minor, for example, changing the term from plural to singular.

Table D-1. New Army terms

Army positive control	enemy	intelligence, surveil-lance, and reconnais-sance synchroni-zation	movement and maneuver warfighting function
Army procedural control	fires warfighting function		
civil support[1]	forward operating base	intelligence warfighting function	operational theme
combat power (Army)			protection warfighting function
command and control warfare	graphic control measure	irregular warfare	situational awareness
command and control warfighting function	influence[2]	knowledge management	stability mechanism
compel	information engagement	landpower	support (Army)
defeat mechanism	information protection	line of effort[3]	supporter
disintegrate	intelligence, surveil-lance, and reconnais-sance integration	line of operations (Army)	sustainment warfighting function
dislocate			unassigned area
			warfighting function

Notes:
[1] Replaces support operations and uses the joint definition with Army primary tasks.
[2] Adds a second definition to an existing term.
[3] Replaces logical line of operations.

Table D-2. Modified Army definitions

assessment (Army)	culminating point (Army)	intelligence, surveillance, and reconnaissance	preparation
battle command			running estimate
close combat	decisive operation	isolate[3]	situational understanding
combined arms	defensive operations	main effort	
commander's intent (Army)	destroy[3]	mission command	stability operations[4]
	essential element of friendly information	mission orders	supporting distance
commander's visualization		neutral	supporting range
	force tailoring	offensive operations	sustaining operation
common operational picture	full spectrum operations	operations process	task-organizing[5]
		phase (Army/Marine Corps)	tempo (Army/Marine Corps)
concept of operations (Army)	information management		urban operations
control (Army)[1, 2]	initiative (individual)	planning	
control measure	initiative (operational)		

Notes:
[1] New definition for use in command and control context.
[2] Added second definition for use as a stability mechanism.
[3] New definition for use in operational art context.
[4] The Army uses the joint definition and assigns Army-specific tasks to this element of full spectrum operations.
[5] Army definition added to joint definition as an addendum.

Table D-3. Rescinded Army definitions

agility[1]	combat service support[2]	offensive information operations (Army)	rear area
assigned forces[1]			subordinates' initiative[5]
asymmetry[1]	combat support[2]	operational fires[1]	
battlefield organization	deep area	operational framework	support operations[6]
	defensive information operations (Army)		versatility[1]
battlespace		operational picture	logical lines of operations[7]
close area	force protection (Army)[3]	protection (Army)[4]	
combat arms			

Notes:
[1] Army doctrine will follow joint definitions and common English usage.
[2] Army doctrine will not use this term; joint doctrine will continue to use this term.
[3] Activities incorporated into the protection warfighting function.
[4] Replaced by protection warfighting function.
[5] Replaced by individual initiative.
[6] Replaced by civil support operations.
[7] Replaced by lines of effort.

Source Notes

These are the sources used, quoted, or paraphrased in this publication. They are listed by page number. Where material appears in a paragraph, both page and paragraph number are listed.

vii   "Stability operations are a core...": DODD 3000.05, *Military Support for Stability, Security, Transition, and Reconstruction (SSTR) Operations* (Washington, DC: Department of Defense, 28 Nov 2005), paragraph 4.1 (page 2).

1-1   "Unfortunately, the dangers and challenges...": Gates, Robert M. (Secretary of Defense), "Landon Lecture," speech given at Kansas State University, (Manhattan, KS, 26 Nov 2007). <http://www.defenselink.mil/speeches/speech.aspx?speechid=1199> (accessed 10 Jan 2008).

1-15   "War is thus an act...": Carl von Clausewitz, *On War*, Michael Howard and Peter Paret, eds. (Princeton: Princeton University Press, 1984), 75 (hereafter cited as Clausewitz).

1-17   "Everything in war...": Clausewitz, 119, 120.

1-20   para 1-87. AR 350-1, *Army Training and Leader Development* (Washington DC: Headquarters, Department of the Army, 3 Aug 2007), 80–81. Paragraph 4-18b establishes the Soldier's Rules. < http://www.army.mil/usapa/epubs/pdf/r350_1.pdf > (accessed 10 Jan 2008).

1-20   para 1-88. Field Marshal Rommel's observation taken from Peter G. Tsouras, ed., *The Greenhill Dictionary of Military Quotations* (London: Greenhill, 2000), 186b.

6-1   "The first, the supreme,...": Clausewitz, 88–89.

6-8   para 6-35. "the hub of all power...": Clausewitz, 595–596.

6-15   "The third factor,...": Clausewitz, 569.

6-19   "In all great actions...": Arthur Wellesley, later 1st Duke of Wellington, in a letter to his brother, the governor general of India, quoted in Arthur Bryant, *The Great Duke or the Invincible General* (New York: William Morrow & Company, Inc., 1972), 57.

7-1   "Be first with the truth.": Multinational Corps–Iraq, Memorandum, Subject: Counterinsurgency Guidance (Baghdad: Headquarters, Multinational Corps–Iraq, undated [estimated 2007], 3).

7-1   paras 7-3–7-4. Vignette based on the personal experiences of the brigade information operations officer (S-7) for 3rd Brigade Combat Team, 10th Mountain Division.

This page intentionally left blank.

Glossary

The glossary lists acronyms and terms with Army, multi-Service, or joint definitions, and other selected terms. Where Army and joint definitions are different, *(Army)* follows the term. Terms for which FM 3-0 is the proponent manual (the authority) are marked with an asterisk (*). The proponent manual for other terms is listed in parentheses after the definition. Terms for which the Army and Marine Corps have agreed on a common definition are followed by *(Army-Marine Corps)*.

SECTION I – ACRONYMS AND ABBREVIATIONS

ADCON	administrative control
AR	Army regulation
ARFOR	*See* ARFOR under terms.
ASCC	Army Service component command
BCT	brigade combat team
CCIR	commander's critical information requirement
CJCSI	Chairman of the Joint Chiefs of Staff instruction
CJCSM	Chairman of the Joint Chiefs of Staff manual
COCOM	combatant command (command authority)
DA	Department of the Army
DOD	Department of Defense
DODD	Department of Defense directive
EEFI	essential element of friendly information
FFIR	friendly force information requirement
FM	field manual
FMI	field manual-interim
ISR	intelligence, surveillance, and reconnaissance
JP	joint publication
MDMP	military decisonmaking process
METT-TC	*See* METT-TC under terms.
NATO	North Atlantic Treaty Organization
OPCON	operational control
PIR	priority intelligence requirement
PMESII-PT	*See* PMESII-PT under terms.
TACON	tactical control
U.S.	United States

SECTION II – TERMS

adversary

(joint) A party acknowledged as potentially hostile to a friendly party and against which the use of force may be envisaged. (JP 3-0)

alliance

(joint) The relationship that results from a formal agreement (for example, a treaty) between two or more nations for broad, long-term objectives that further the common interests of the members. (JP 3-0)

area of influence

(joint) A geographical area wherein a commander is directly capable of influencing operations by maneuver or fire support systems normally under the commander's command or control. (JP 1-02)

area of interest

(joint) That area of concern to the commander, including the area of influence, areas adjacent thereto, and extending into enemy territory to the objectives of current or planned operations. This area also includes areas occupied by enemy forces who could jeopardize the accomplishment of the mission. (JP 2-03)

area of operations

(joint) An operational area defined by the joint force commander for land and maritime forces. Areas of operations do not typically encompass the entire operational area of the joint force commander, but should be large enough for component commanders to accomplish their missions and protect their forces. (JP 3-0)

*ARFOR

The Army Service component headquarters for a joint task force or a joint and multinational force.

*Army positive control

A technique of regulating forces that involves commanders and leaders actively assessing, deciding, and directing them.

*Army procedural control

A technique of regulating forces that relies on a combination of orders, regulations, policies, and doctrine (including tactics, techniques, and procedures).

*assessment

(Army) The continuous monitoring and evaluation of the current situation, particularly the enemy, and progress of an operation.

*battle

A set of related engagements that lasts longer and involves larger forces than an engagement.

*battle command

The art and science of understanding, visualizing, describing, directing, leading, and assessing forces to impose the commander's will on a hostile, thinking, and adaptive enemy. Battle command applies leadership to translate decisions into actions—by synchronizing forces and warfighting functions in time, space, and purpose—to accomplish missions.

branch

(joint) The contingency options built into the base plan. A branch is used for changing the mission, orientation, or direction of movement of a force to aid success of the operation based on anticipated events, opportunities, or disruptions caused by enemy actions and reactions. (JP 5-0)

campaign

(joint) A series of related major operations aimed at achieving strategic and operational objectives within a given time and space. (JP 5-0)

center of gravity

    (joint) The source of power that provides moral or physical strength, freedom of action, or will to act. (JP 3-0)

civil considerations

    The influence of manmade infrastructure, civilian institutions, and attitudes and activities of the civilian leaders, populations, and organizations within an area of operations on the conduct of military operations. (FM 6-0)

civil support

    (joint) Department of Defense support to U.S. civil authorities for domestic emergencies, and for designated law enforcement and other activities. (JP 1-02)

*close combat

    Warfare carried out on land in a direct-fire fight, supported by direct, indirect, and air-delivered fires.

coalition

    (joint) An ad hoc arrangement between two or more nations for common action. (JP 5-0)

coalition action

    (joint) Multinational action outside the bounds of established alliances, usually for single occasions or longer cooperation in a narrow sector of common interest. (JP 5-0)

combatant command (command authority)

    (joint) Nontransferable command authority established by title 10 ("Armed Forces"), United States Code, section 164, exercised only by commanders of unified or specified combatant commands unless otherwise directed by the President or the Secretary of Defense. Combatant command (command authority) cannot be delegated and is the authority of a combatant commander to perform those functions of command over assigned forces involving organizing and employing commands and forces, assigning tasks, designating objectives, and giving authoritative direction over all aspects of military operations, joint training, and logistics necessary to accomplish the missions assigned to the command. Combatant command (command authority) should be exercised through the commanders of subordinate organizations. Normally this authority is exercised through subordinate joint force commanders and Service and/or functional component commanders. Combatant command (command authority) provides full authority to organize and employ commands and forces as the combatant commander considers necessary to accomplish assigned missions. Operational control is inherent in combatant command (command authority). (JP 1)

combat camera

    (joint) The acquisition and utilization of still and motion imagery in support of combat, information, humanitarian, special force, intelligence, reconnaissance, engineering, legal, public affairs, and other operations involving the Military Services. (JP 3-61)

combat information

    (joint) Unevaluated data, gathered by or provided directly to the tactical commander which, due to its highly perishable nature or the criticality of the situation, cannot be processed into tactical intelligence in time to satisfy the user's tactical intelligence requirements. (JP 1-02)

*combat power

    (Army) The total means of destructive, constructive, and information capabilities that a military unit/formation can apply at a given time. Army forces generate combat power by converting potential into effective action.

*combined arms

    The synchronized and simultaneous application of the elements of combat power—to achieve an effect greater than if each element of combat power was used separately or sequentially.

*command

(joint) The authority that a commander in the armed forces lawfully exercises over subordinates by virtue of rank or assignment. Command includes the authority and responsibility for effectively using available resources and for planning the employment of, organizing, directing, coordinating, and controlling military forces for the accomplishment of assigned missions. It also includes responsibility for health, welfare, morale, and discipline of assigned personnel. (JP 1)

command and control

(Army) The exercise of authority and direction by a properly designated commander over assigned and attached forces in the accomplishment of a mission. Commanders perform command and control functions through a command and control system. (FM 6-0)

command and control system

(Army) The arrangement of personnel, information management, procedures, and equipment and facilities essential for the commander to conduct operations. (FM 6-0)

*command and control warfare

The integrated use of physical attack, electronic warfare, and computer network operations, supported by intelligence, to degrade, destroy, and exploit an enemy's or adversary's command and control system or to deny information to it.

*command and control warfighting function

The related tasks and systems that support commanders in exercising authority and direction.

commander's critical information requirement

(joint) An information requirement identified by the commander as being critical to facilitating timely decisionmaking. The two key elements are friendly force information requirements and priority intelligence requirements. (JP 3-0)

*commander's intent

(Army) A clear, concise statement of what the force must do and the conditions the force must establish with respect to the enemy, terrain, and civil considerations that represent the desired end state.

*commander's visualization

The mental process of developing situational understanding, determining a desired end state, and envisioning the broad sequence of events by which the force will achieve that end state.

*common operational picture

(Army) A single display of relevant information within a commander's area of interest tailored to the user's requirements and based on common data and information shared by more than one command.

*compel

To use, or threaten to use, lethal force to establish control and dominance, effect behavioral change, or enforce compliance with mandates, agreements, or civil authority.

*concept of operations

(Army) A statement that directs the manner in which subordinate units cooperate to accomplish the mission and establishes the sequence of actions the force will use to achieve the end state. It is normally expressed in terms of decisive, shaping, and sustaining operations.

conduct

To perform the activities of the operations process: planning, preparing, executing, and continuously assessing. (FM 6-0)

consequence management

(joint) Actions taken to maintain or restore essential services and manage and mitigate problems resulting from disasters and catastrophes, including natural, man-made, or terrorist incidents. (JP 1-02)

contractor

A person or business that provides products or services for monetary compensation. A contractor furnishes supplies and services or performs work at a certain price or rate based on the terms of a contract. (FM 3-100.21)

*control

(Army) *1. In the context of command and control, the regulation of forces and warfighting functions to accomplish the mission in accordance with the commander's intent. (FM 3-0) 2. A tactical mission task that requires the commander to maintain physical influence over a specified area to prevent its use by an enemy. (FM 3-90) 3. An action taken to eliminate a hazard or reduce its risk. (FM 5-19) *4. In the context of stability mechanisms, to impose civil order. (FM 3-0) [See JP 1-02 for joint definitions.]

*control measure

A means of regulating forces or warfighting functions.

conventional forces

(joint) 1. Those forces capable of conducting operations using nonnuclear weapons. 2. Those forces other than designated special operations forces. (JP 3-05)

counterdrug activities

(joint) Those measures taken to detect, interdict, disrupt, or curtail any activity that is reasonably related to illicit drug trafficking. This includes, but is not limited to, measures taken to detect, interdict, disrupt, or curtail activities related to substances, materiel, weapons, or resources used to finance, support, secure, cultivate, process, or transport illegal drugs. (JP 3-07.4)

counterinsurgency

(joint) Those military, paramilitary, political, economic, psychological, and civic actions taken by a government to defeat insurgency. (JP 1-02)

*culminating point

(Army) That point in time and space at which a force no longer possesses the capability to continue its current form of operations.

*decisive operation

The operation that directly accomplishes the mission. It determines the outcome of a major operation, battle, or engagement. The decisive operation is the focal point around which commanders design the entire operation.

decisive point

(joint) A geographic place, specific key event, critical factor, or function that, when acted upon, allows commanders to gain a marked advantage over an adversary or contribute materially to achieving suc–cess. (JP 3-0) [Note: In this context, adversary also refers to enemies.]

*defeat mechanism

The method through which friendly forces accomplish their mission against enemy opposition.

*defensive operations

Combat operations conducted to defeat an enemy attack, gain time, economize forces, and develop conditions favorable for offensive or stability operations.

*depth

(Army) The extension of operations in time, space, and resources.

*destroy

In the context of defeat mechanisms, to apply lethal combat power on an enemy capability so that it can no longer perform any function and cannot be restored to a usable condition without being entirely rebuilt.

*direct approach

An operational approach that attacks the enemy's center of gravity or principal strength by applying combat power directly against it.

*disintegrate

To disrupt the enemy's command and control system, degrading the ability to conduct operations while leading to a rapid collapse of enemy's capabilities or will to fight.

*dislocate

To employ forces to obtain significant positional advantage, rendering the enemy's dispositions less valuable, perhaps even irrelevant.

end state

(joint) The set of required conditions that defines achievement of the commander's objectives. (JP 3-0)

*enemy

A party identified as hostile against which the use of force is authorized.

engagement

(joint) A tactical conflict, usually between opposing, lower echelon maneuver forces. (JP 1-02)

*essential element of friendly information

(Army) A critical aspect of a friendly operation that, if known by the enemy, would subsequently compromise, lead to failure, or limit success of the operation, and therefore should be protected from enemy detection.

*execution

Putting a plan into action by applying combat power to accomplish the mission and using situational understanding to assess progress and make execution and adjustment decisions.

*exterior lines

A force operates on exterior lines when its operations converge on the enemy.

*fires warfighting function

The related tasks and systems that provide collective and coordinated Army indirect fires, joint fires, and command and control warfare, including nonlethal fires, through the targeting process.

*force tailoring

The process of determining the right mix of forces and the sequence of their deployment in support of a joint force commander.

foreign humanitarian assistance

(joint) Programs conducted to relieve or reduce the results of natural or man-made disasters or other endemic conditions such as human pain, disease, hunger, or privation that might present a serious threat to life or that can result in great damage to or loss of property. Foreign humanitarian assistance provided by U.S. forces is limited in scope and duration. The foreign assistance provided is designed to supplement or complement the efforts of the host-nation civil authorities or agencies that may have the primary responsibility for providing foreign humanitarian assistance. Foreign humanitarian assistance operations are those conducted outside the United States, its territories, and possessions. (JP 3-33)

*forward operating base

An area used to support tactical operations without establishing full support facilities.

friendly force information requirement

(joint) Information the commander and staff need to understand the status of friendly force and supporting capabilities. (JP 3-0)

*full spectrum operations

The Army's operational concept: Army forces combine offensive, defensive, and stability or civil support operations simultaneously as part of an interdependent joint force to seize, retain, and exploit the initiative, accepting prudent risk to create opportunities to achieve decisive results. They employ synchronized action—lethal and nonlethal—proportional to the mission and informed by a thorough understanding of all variables of the operational environment. Mission command that conveys intent and an appreciation of all aspects of the situation guides the adaptive use of Army forces.

general war

(joint) Armed conflict between major powers in which the total resources of the belligerents are employed, and the national survival of a major belligerent is in jeopardy. (JP 1-02)

*graphic control measure

A symbol used on maps and displays to regulate forces and warfighting functions.

*indirect approach

An operational approach that attacks the enemy's center of gravity by applying combat power against a series of decisive points while avoiding enemy strength.

individual initiative

See initiative (individual).

*influence

In the context of stability mechanisms, to alter the opinions and attitudes of a civilian population through information engagement, presence, and conduct.

*information engagement

The integrated employment of public affairs to inform U.S. and friendly audiences; psychological operations, combat camera, U.S. Government strategic communication and defense support to public diplomacy, and other means necessary to influence foreign audiences; and, leader and Soldier engagements to support both efforts.

information environment

(joint) The aggregate of individuals, organizations, and systems that collect, process, disseminate, or act on information. (JP 3-13)

*information management

(Army) The science of using procedures and information systems to collect, process, store, display, disseminate, and protect knowledge products, data, and information.

information operations

(joint) The integrated employment of the core capabilities of electronic warfare, computer network operations, psychological operations, military deception, and operations security, in concert with specified supporting and related capabilities, to influence, disrupt, corrupt, or usurp adversarial human and automated decisionmaking while protecting our own. (JP 3-13)

*information protection

Active or passive measures that protect and defend friendly information and information systems to ensure timely, accurate, and relevant friendly information. It denies enemies, adversaries, and others the opportunity to exploit friendly information and information systems for their own purposes.

information superiority

(joint) The operational advantage derived from the ability to collect, process, and disseminate an uninterrupted flow of information while exploiting or denying an adversary's ability to do the same. (JP 3-13) [Note: In this context, adversary also refers to enemies.]

*information system

(Army) Equipment and facilities that collect, process, store, display, and disseminate information. This includes computers—hardware and software—and communications, as well as policies and procedures for their use.

*initiative (individual)

The willingness to act in the absence of orders, when existing orders no longer fit the situation, or when unforeseen opportunities or threats arise.

*initiative (operational)

The setting or dictating the terms of action throughout an operation.

insurgency

(joint) An organized movement aimed at the overthrow of a constituted government through use of subversion and armed conflict. (JP 1-02)

intelligence

(joint) The product resulting from the collection, processing, integration, evaluation, analysis, and interpretation of available information concerning foreign nations, hostile or potentially hostile forces or elements, or areas of actual or potential operations. The term is also applied to the activity which results in the product and to the organizations engaged in such activity. (JP 2-0)

*intelligence, surveillance, and reconnaissance

(Army) An activity that synchronizes and integrates the planning and operation of sensors, assets, and processing, exploitation, and dissemination systems in direct support of current and future operations. This is an integrated intelligence and operations function. For Army forces, this activity is a combined arms operation that focuses on priority intelligence requirements while answering the commander's critical information requirements.

*intelligence, surveillance, and reconnaissance integration

The task of assigning and controlling a unit's intelligence, surveillance, and reconnaissance assets (in terms of space, time, and purpose) to collect and report information as a concerted and integrated portion of operation plans and orders.

*intelligence, surveillance, and reconnaissance synchronization

The task that accomplishes the following: analyzes information requirements and intelligence gaps; evaluates available assets internal and external to the organization; determines gaps in the use of those assets; recommends intelligence, surveillance, and reconnaissance assets controlled by the organization to collect on the commander's critical information requirements; and submits requests for information for adjacent and higher collection support.

*intelligence warfighting function

The related tasks and systems that facilitate understanding of the operational environment, enemy, terrain, and civil considerations.

interagency coordination

(joint) Within the context of Department of Defense involvement, the coordination that occurs between elements of Department of Defense and engaged U.S. Government agencies for the purpose of achieving an objective. (JP 3-0)

*interior lines

A force operates on interior lines when its operations diverge from a central point.

irregular forces

(joint) Armed individuals or groups who are not members of the regular armed forces, police, or other internal security forces. (JP 1-02)

*irregular warfare

A violent struggle among state and nonstate actors for legitimacy and influence over a population.

*isolate

> In the context of defeat mechanisms, to deny an enemy or adversary access to capabilities that enable the exercise of coercion, influence, potential advantage, and freedom of action.

joint combined exchange training

> (joint) A program conducted overseas to fulfill U.S. forces training requirements and at the same time exchange the sharing of skills between U.S. forces and host-nation counterparts. Training activities are designed to improve U.S. and host-nation capabilities. (JP 3-05)

*knowledge management

> The art of creating, organizing, applying, and transferring knowledge to facilitate situational understanding and decisionmaking. Knowledge management supports improving organizational learning, innovation, and performance. Knowledge management processes ensure that knowledge products and services are relevant, accurate, timely, and useable to commanders and decisionmakers.

*landpower

> The ability—by threat, force, or occupation—to gain, sustain, and exploit control over land, resources, and people.

law of war

> (joint) That part of international law that regulates the conduct of armed hostilities. (JP 1-02)

leadership

> The process of influencing people by providing purpose, direction, and motivation, while operating to accomplish the mission and improving the organization. (FM 6-22)

*line of effort

> A line that links multiple tasks and missions using the logic of purpose—cause and effect—to focus efforts toward establishing operational and strategic conditions.

*line of operations

> (Army) A line that defines the directional orientation of a force in time and space in relation to the enemy and links the force with its base of operations and objectives.

*main effort

> The designated subordinate unit whose mission at a given point in time is most critical to overall mission success. It is usually weighted with the preponderance of combat power.

major operation

> (joint) A series of tactical actions (battles, engagements, strikes) conducted by combat forces of a single or several Services, coordinated in time and place, to achieve strategic or operational objectives in an operational area. These actions are conducted simultaneously or sequentially in accordance with a common plan and are controlled by a single commander. For noncombat operations, a reference to the relative size and scope of a military operation. (JP 3-0)

maneuver

> (joint) Employment of forces in the operational area through movement in combination with fires to achieve a position of advantage in respect to the enemy in order to accomplish the mission. (JP 3-0)

measure of effectiveness

> (joint) A criterion used to assess changes in system behavior, capability, or operational environment that is tied to measuring the attainment of an end state, achievement of an objective, or creation of an effect. (JP 3-0)

measure of performance

> (joint) A criterion used to assess friendly actions that is tied to measuring task accomplishment. (JP 3-0)

METT-TC

A memory aid used in two contexts: 1. In the context of information management, the major subject categories into which relevant information is grouped for military operations: mission, enemy, terrain and weather, troops and support available, time available, civil considerations. (FM 6-0) 2. In the context of tactics, major variables considered during mission analysis (mission variables). (FM 3-90)

mission

(joint) The task, together with the purpose, that clearly indicates the action to be taken and the reason therefor. (JP 1-02)

*mission command

The conduct of military operations through decentralized execution based on mission orders. Successful mission command demands that subordinate leaders at all echelons exercise disciplined initiative, acting aggressively and independently to accomplish the mission within the commander's intent.

*mission orders

A technique for developing orders that emphasizes to subordinates the results to be attained, not how they are to achieve them. It provides maximum freedom of action in determining how to best accomplish assigned missions.

*movement and maneuver warfighing function

The related tasks and systems that move forces to achieve a position of advantage in relation to the enemy. Direct fire is inherent in maneuver, as is close combat.

multinational operations

(joint) A collective term to describe military actions conducted by forces of two or more nations, usually undertaken within the structure of a coalition or alliance. (JP 3-16)

*neutral

(Army) A party identified as neither supporting nor opposing friendly or enemy forces.

noncombatant evacuation operations

(joint) Operations directed by the Department of State or other appropriate authority, in conjunction with the Department of Defense, whereby noncombatants are evacuated from foreign countries when their lives are endangered by war, civil unrest, or natural disaster to safe havens or to the United States. (JP 3-0)

*offensive operations

Combat operations conducted to defeat and destroy enemy forces and seize terrain, resources, and population centers. They impose the commander's will on the enemy.

*operational approach

The manner in which a commander contends with a center of gravity.

operational area

(joint) An overarching term encompassing more descriptive terms for geographic areas in which military operations are conducted. Operational areas include, but are not limited to, such descriptors as area of responsibility, theater of war, theater of operations, joint operations area, amphibious objective area, joint special operations area, and area of operations. (JP 5-0)

operational art

(joint) The application of creative imagination by commanders and staffs—supported by their skill, knowledge, and experience—to design strategies, campaigns, and major operations and organize and employ military forces. Operational art integrates ends, ways, and means across the levels of war. (JP 3-0)

operational concept

See full spectrum operations.

operational control

> (joint) Command authority that may be exercised by commanders at any echelon at or below the level of combatant command. Operational control is inherent in combatant command (command authority) and may be delegated within the command. Operational control is the authority to perform those functions of command over subordinate forces involving organizing and employing commands and forces, assigning tasks, designating objectives, and giving authoritative direction necessary to accomplish the mission. Operational control includes authoritative direction over all aspects of military operations and joint training necessary to accomplish missions assigned to the command. Operational control should be exercised through the commanders of subordinate organizations. Normally this authority is exercised through subordinate joint force commanders and Service and/or functional component commanders. Operational control normally provides full authority to organize commands and forces and to employ those forces as the commander in operational control considers necessary to accomplish assigned missions; it does not, in and of itself, include authoritative direction for logistics or matters of administration, discipline, internal organization, or unit training. (JP 1)

operational environment

> (joint) A composite of the conditions, circumstances, and influences that affect the employment of capabilities and bear on the decisions of the commander. (JP 3-0)

operational initiative

> *See* initiative (operational).

*operational pause

> (Army) A deliberate halt taken to extend operational reach or prevent culmination.

operational reach

> (joint) The distance and duration across which a unit can successfully employ military capabilities. (JP 3-0)

*operational theme

> The character of the dominant major operation being conducted at any time within a land force commander's area of operations. The operational theme helps convey the nature of the major operation to the force to facilitate common understanding of how the commander broadly intends to operate.

*operations process

> The major command and control activities performed during operations: planning, preparing, executing, and continuously assessing the operation. The commander drives the operations process.

peace building

> (joint) Stability actions, predominately diplomatic and economic, that strengthen and rebuild governmental infrastructure and institutions in order to avoid a relapse into conflict. (JP 3-0)

peacekeeping

> (joint) Military operations undertaken with the consent of all major parties to a dispute, designed to monitor and facilitate implementation of an agreement (cease fire, truce, or other such agreement) and support diplomatic efforts to reach a long-term political settlement. (JP 3-07.3)

peacemaking

> (joint) The process of diplomacy, mediation, negotiation, or other forms of peaceful settlements that arranges an end to a dispute and resolves issues that led to it. (JP 3-0)

peace operations

> (joint) A broad term that encompasses multiagency and multinational crisis response and limited contingency operations involving all instruments of national power with military missions to contain conflict, redress the peace, and shape the environment to support reconciliation and rebuilding and facilitate the transition to legitimate governance. Peace operations include peacekeeping, peace enforcement, peacemaking, peace building, and conflict prevention efforts. (JP 3-07.3)

*peacetime military engagement

All military activities that involve other nations and are intended to shape the security environment in peacetime. It includes programs and exercises that the United States military conducts with other nations to shape the international environment, improve mutual understanding, and improve interoperability with treaty partners or potential coalition partners. Peacetime military engagement activities are designed to support a combatant commander's objectives within the theater security cooperation plan.

*phase

(Army/Marine Corps) A planning and execution tool used to divide an operation in duration or activity. A change in phase usually involves a change of mission, task organization, or rules of engagement. Phasing helps in planning and controlling and may be indicated by time, distance, terrain, or an event.

plan

A design for a future or anticipated operation. (FM 5-0)

*planning

The process by which commanders (and the staff, if available) translate the commander's visualization into a specific course of action for preparation and execution, focusing on the expected results.

*PMESII-PT

A memory aid for the varibles used to describe the operational environment: political, military, economic, social, information, infrastructure, physical environment, time (operational variables).

*preparation

Activities performed by units to improve their ability to execute an operation. Preparation includes, but is not limited to, plan refinement; rehearsals; intelligence, surveillance, and reconnaissance; coordination; inspections; and movement.

priority intelligence requirement

(joint) An intelligence requirement, stated as a priority for intelligence support, that the commander and staff need to understand the adversary or the operational environment. (JP 2-0) [Note: In this context, adversary also refers to enemies.]

*protection warfighting function

The related tasks and systems that preserve the force so the commander can apply maximum combat power.

raid

(joint) An operation to temporarily seize an area in order to secure information, confuse an adversary, capture personnel or equipment, or to destroy a capability. It ends with a planned withdrawal upon completion of the assigned mission. (JP 3-0) [Note: In this context, adversary also refers to enemies.]

recovery operations

(joint) Operations conducted to search for, locate, identify, recover, and return isolated personnel, human remains, sensitive equipment, or items critical to national security. (JP 3-50)

*relevant information

All information of importance to commanders and staffs in the exercise of command and control.

rules of engagement

(joint) Directives issued to guide United States forces on the use of force during various operations. These directives may take the form of execute orders, deployment orders, memoranda of agreement, or plans. (JP 1-02)

*running estimate

A staff section's continuous assessment of current and future operations to determine if the current operation is proceeding according to the commander's intent and if future operations are supportable.

sanction enforcement

> (joint) Operations that employ coercive measures to interdict the movement of certain types of designated items into or out of a nation or specified area. (JP 3-0)

sequel

> (joint) In a campaign, a major operation that follows the current major operation. In a single major operation, a sequel is the next phase. Plans for a sequel are based on the possible outcomes (success, stalemate, or defeat) associated with the current operation. (JP 5-0)

*shaping operation

> An operation at any echelon that creates and preserves conditions for the success of the decisive operation.

show of force

> (joint) An operation designed to demonstrate U.S. resolve that involves increased visibility of U.S. deployed forces in an attempt to defuse a specific situation that, if allowed to continue, may be detrimental to U.S. interests or national objectives. (JP 3-0)

*situational awareness

> Immediate knowledge of the conditions of the operation, constrained geographically and in time.

*situational understanding

> The product of applying analysis and judgment to relevant information to determine the relationships among the mission variables to facilitate decisionmaking.

*stability mechanism

> The primary method through which friendly forces affect civilians in order to attain conditions that support establishing a lasting, stable peace.

stability operations

> (joint) An overarching term encompassing various military missions, tasks, and activities conducted outside the United States in coordination with other instruments of national power to maintain or reestablish a safe and secure environment, provide essential governmental services, emergency infrastructure reconstruction, and humanitarian relief. (JP 3-0)

strike

> (joint) An attack to damage or destroy an objective or a capability. (JP 3-0)

*support

> (joint) The action of a force that aids, protects, complements, or sustains another force in accordance with a directive requiring such action. (JP 1) (Army) *In the context of stability mechanisms, to establish, reinforce, or set the conditions necessary for the other instruments of national power to function effectively.

*supporter

> A party who sympathizes with friendly forces and who may or may not provide material assistance to them.

*supporting distance

> The distance between two units that can be traveled in time for one to come to the aid of the other and prevent its defeat by an enemy or ensure it regains control of a civil situation.

*supporting range

> The distance one unit may be geographically separated from a second unit yet remain within the maximum range of the second unit's weapons systems.

*sustaining operation

> An operation at any echelon that enables the decisive operation or shaping operations by generating and maintaining combat power.

*sustainment warfighting function

    The related tasks and systems that provide support and services to ensure freedom of action, extend operational reach, and prolong endurance.

synchronization

    (joint) The arrangement of military actions in time, space, and purpose to produce maximum relative combat power at a decisive place and time. (JP 2-0)

system

    (joint) A functionally, physically, and/or behaviorally related group of regularly interacting or interdependent elements; that group of elements forming a unified whole. (JP 3-0)

tactical combat force

    (joint) A combat unit, with appropriate combat support and combat service support assets, that is assigned the mission of defeating level III threats. (JP 3-10)

tactical control

    (joint) Command authority over assigned or attached forces or commands, or military capability or forces made available for tasking, that is limited to the detailed direction and control of movements or maneuvers within the operational area necessary to accomplish missions or tasks assigned. Tactical control is inherent in operational control. Tactical control may be delegated to, and exercised at any level at or below the level of combatant command. Tactical control provides sufficient authority for controlling and directing the application of force or tactical use of combat support assets within the assigned mission or task. (JP 1)

tactics

    (joint) The employment and ordered arrangement of forces in relation to each other. (CJCSI 5120.02A)

*task organization

    (Army) A temporary grouping of forces designed to accomplish a particular mission.

*task-organizing

    (Army) The act of designing an operating force, support staff, or logistic package of specific size and composition to meet a unique task or mission. Characteristics to examine when task-organizing the force include, but are not limited to: training, experience, equipage, sustainability, operating environment, enemy threat, and mobility. For Army forces, it includes allocating available assets to subordinate commanders and establishing their command and support relationships.

*tempo

    (Army/Marine Corps) The relative speed and rhythm of military operations over time with respect to the enemy.

terrorism

    (joint) The calculated use of unlawful violence or threat of unlawful violence to inculcate fear; intended to coerce or to intimidate governments or societies in the pursuit of goals that are generally political, religious, or ideological. (JP 3-07.2)

*unassigned area

    The area between noncontiguous areas of operations or beyond contiguous areas of operations. The higher headquarters is responsible for controlling unassigned areas within its area of operations.

unconventional warfare

    (joint) A broad spectrum of military and paramilitary operations, normally of long duration, predominantly conducted through, with, or by indigenous or surrogate forces who are organized, trained, equipped, supported, and directed in varying degrees by an external source. It includes, but is not limited to, guerrilla warfare, subversion, sabotage, intelligence activities, and unconventional assisted recovery. (JP 3-05)

unified action

(joint) The synchronization, coordination, and/or integration of the activities of governmental and nongovernmental entities with military operations to achieve unity of effort. (JP 1)

*urban operation

A military operation conducted where man-made construction and high population density are the dominant features.

*warfighting function

A group of tasks and systems (people, organizations, information, and processes) united by a common purpose that commanders use to accomplish missions and training objectives.

This page intentionally left blank.

References

Field manuals and selected joint publications are listed by new number followed by old number.

REQUIRED PUBLICATIONS

These documents must be available to intended users of this publication.

FM 1-02 (101-5-1). *Operational Terms and Graphics.* 21 September 2004.

JP 1-02. *Department of Defense Dictionary of Military and Associated Terms.* 12 April 2001.

RELATED PUBLICATIONS

These sources contain relevant supplemental information.

JOINT AND DEPARTMENT OF DEFENSE PUBLICATIONS

Most joint publications are available online: http://www.dtic.mil/doctrine/jpcapstonepubs.htm.

CJCSI 5120.02A. *Joint Doctrine Development System.* 31 March 2007.

DODD 3000.05. *Military Support for Stability, Security, Transition, and Reconstruction (SSTR) Operations.* 28 November 2005.

DODD 5100.1. *Functions of the Department of Defense and Its Major Components.* 1 August 2002.

JP 1. *Doctrine for the Armed Forces of the United States.* 14 May 2007.

JP 2-0. *Joint Intelligence.* 22 June 2007.

JP 2-01. *Joint and National Intelligence Support to Military Operations.* 7 October 2004.

JP 2-03. *Geospatial Intelligence Support to Joint Operations.* 22 March 2007.

JP 3-0. *Joint Operations.* 17 September 2006.

JP 3-05. *Doctrine for Joint Special Operations.* 17 December 2003.

JP 3-06. *Doctrine for Joint Urban Operations.* 16 September 2002.

JP 3-07.1. *Joint Tactics, Techniques, and Procedures for Foreign Internal Defense (FID).* 30 April 2004.

JP 3-07.2. *Antiterrorism.* 14 April 2006.

JP 3-07.3. *Peace Operations.* 17 October 2007.

JP 3-07.4. *Joint Counterdrug Operations.* 13 June 2007.

JP 3-07.6. *Joint Tactics, Techniques, and Procedures for Foreign Humanitarian Assistance.* 15 August 2001.

JP 3-08. *Interagency, Intergovernmental Organization, and Nongovernmental Organization Coordination During Joint Operations* (2 volumes). 17 March 2006.

JP 3-09.1. *Joint Tactics, Techniques, and Procedures for Laser Designation Operations.* 28 May 1999.

JP 3-10. *Joint Security Operations in Theater.* 1 August 2006.

JP 3-13. *Information Operations.* 13 February 2006.

JP 3-13.1. *Electronic Warfare.* 25 January 2007.

JP 3-13.3 (JP 3-54). *Operations Security.* 29 June 2006.

JP 3-13.4 (JP 3-58). *Military Deception.* 13 July 2006.

JP 3-16. *Multinational Operations.* 7 March 2007.

JP 3-18. *Joint Doctrine for Forcible Entry Operations.* 16 July 2001.

JP 3-28. *Civil Support.* 14 September 2007.

JP 3-33. *Joint Task Force Headquarters.* 16 February 2007.

JP 3-35. *Deployment and Redeployment Operations.* 7 May 2007.

JP 3-40. *Joint Doctrine for Combating Weapons of Mass Destruction.* 8 July 2004.

JP 3-41. *Chemical, Biological, Radiological, Nuclear, and High-Yield Explosives Consequence Management.* 2 October 2006.

JP 3-50. *Personnel Recovery.* 5 January 2007.

JP 3-57. *Joint Doctrine for Civil-Military Operations.* 8 February 2001.

JP 3-57.1. *Joint Doctrine for Civil Affairs.* 14 April 2003.

JP 3-61. *Public Affairs.* 9 May 2005.

JP 3-68. *Noncombatant Evacuation Operations.* 22 January 2007.

JP 4-0. *Doctrine for Logistic Support of Joint Operations.* 6 April 2000.

JP 4-05. *Joint Mobilization Planning.* 11 January 2006.

JP 5-0. *Joint Operation Planning.* 26 December 2006.

JP 6-0. *Joint Communications System.* 20 March 2006.

ARMY PUBLICATIONS

Most Army doctrinal publications are available online: https://akocomm.us.army.mil/usapa/doctrine/. Army regulations are produced only in electronic media. Most are available online: https://akocomm.us.army.mil/usapa/epubs/index.html

AR 10-87. *Army Command, Army Service Component Commands, and Direct Reporting Units.* 4 September 2007.

AR 12-1. *Security Assistance, International Logistics, Training, and Technical Assistance Support Policy and Responsibilities.* 24 January 2000.

AR 34-1. *Multinational Force Compatibility.* 6 January 2004.

AR 350-1. *Army Training and Leader Development.* 3 August 2007.

AR 360-1. *The Army Public Affairs Program.* 15 September 2000.

FM 1. *The Army.* 14 June 2005.

FM 2-0 (34-1). *Intelligence.* 17 May 2004.

FM 3-05.202 (31-20-3). *Special Forces Foreign Internal Defense Operations.* 2 February 2007.

FM 3-05.40 (41-10). *Civil Affairs Operations.* 29 September 2006.

FM 3-05.401. *Civil Affairs Tactics, Techniques, and Procedures.* 5 July 2007.

FM 3-06. *Urban Operations.* 26 October 2006.

FM 3-07. *Stability Operations and Support Operations.* 20 February 2003. (When revised, FM 3-07 will be republished as *Stability Operations.*)

FM 3-13 (100-6). *Information Operations: Doctrine, Tactics, Techniques, and Procedures.* 28 November 2003.

FM 3-24. *Counterinsurgency.* 15 December 2006.

FM 3-50.1. *Army Personnel Recovery.* 10 August 2005.

FM 3-61.1. *Public Affairs, Tactics, Techniques and Procedures.* 1 October 2000.

FM 3-90. *Tactics.* 4 July 2001.

FM 3-100.21 (100-21). *Contractors on the Battlefield.* 3 January 2003.

FM 4-0 (100-10). *Combat Service Support.* 29 August 2003.

FM 4-01.011 (55-9, 55-65). *Unit Movement Operations.* 31 October 2002.

FM 5-0 (101-5). *Army Planning and Orders Production.* 20 January 2005.

FM 5-19 (100-14). *Composite Risk Management.* 21 August 2006.

FM 6-0. *Mission Command: Command and Control of Army Forces.* 11 August 2003.

FM 6-20-10. *Tactics, Techniques, and Procedures for the Targeting Process.* 8 May 1996.

FM 6-22 (22-100). *Army Leadership.* 12 October 2006.

FM 7-0 (25-100). *Training the Force.* 22 October 2002.

FM 7-1 (25-101). *Battle Focused Training.* 15 September 2003.

FM 27-10. *The Law of Land Warfare.* 18 July 1956.

FM 46-1. *Public Affairs Operations.* 30 May 1997. (When revised, FM 46-1 will be republished as FM 3-61.)

FM 100-8. *The Army in Multinational Operations.* 24 November 1997. (When revised, FM 100-8 will be republished as FM 3-16.)

FM 100-17-1. *Army Pre-Positioned Afloat Operations.* 27 July 1996.

FM 100-17-2. *Army Pre-Positioned Land.* 16 February 1999.

FMI 3-0.1. *The Modular Force.* 28 January 2008. (The doctrine in FMI 3-0.1 will be incorporated into a revised FM 3-90, FM 3-91, and publications addressing specific echelons and organization types.)

FMI 3-35 (FMs 3-35.4, 100-17, 100-17.-3, 100-17-5). *Army Deployment and Redeployment.* 15 June 2007.

FMI 5-0.1. *The Operations Process.* 31 March 2006.

OTHER PUBLICATIONS

National Military Strategy of the United States of America. Washington, DC: U.S. Government Printing Office. 2005.

WEB SITES

United Nations website (www.un.org).

SOURCES USED

Arthur Bryant. *The Great Duke or the Invincible General.* New York: William Morrow & Company, Inc., 1972.

Clausewitz, Carl von. *On War.* Edited by Michael Howard and Peter Paret. Princeton: Princeton University Press, 1984.

Gates, Robert M. (Secretary of Defense). "Landon Lecture." Speech given at Kansas State University, Manhattan, KS, 26 Nov 2007. <http://www.defenselink.mil/speeches/speech.aspx?speechid=1199> (accessed 10 Jan 2007).

PRESCRIBED FORMS

None

REFERENCED FORMS

None

This page intentionally left blank.

Index

Entries are by paragraph number.

Entries are by paragraph number.

Entries are by paragraph number.

Entries are by paragraph number.

Entries are by paragraph number.

Entries are by paragraph number.

Entries are by paragraph number.

Entries are by paragraph number.

Entries are by paragraph number.

Entries are by paragraph number.

Entries are by paragraph number.

Entries are by paragraph number.

Entries are by paragraph number.

Entries are by paragraph number.

Entries are by paragraph number.

Entries are by paragraph number.

This page is intentionally left blank.

www.ingramcontent.com/pod-product-compliance
Lightning Source LLC
Chambersburg PA
CBHW080500110426
42742CB00017B/2949